LIFE BREAKS IN

LIFE BREAKS IN

(a mood almanack)

MARY CAPPELLO

The University of Chicago Press
Chicago and London

The University of Chicago Press, Chicago 60637
The University of Chicago Press, Ltd., London
© 2016 by Mary Cappello
All rights reserved. Published 2016.
Printed in the United States of America

25 24 23 22 21 20 19 18 17 16   1 2 3 4 5

ISBN-13: 978-0-226-35606-8 (cloth)
ISBN-13: 978-0-226-35623-5 (e-book)
DOI: 10.7208/chicago/9780226356235.001.0001

Library of Congress Cataloging-in-Publication Data

Names: Cappello, Mary, author.
Title: Life breaks in : a mood almanack / Mary Cappello.
Description: Chicago : The University of Chicago Press, 2016. | Includes bibliographical references and
    index.
Identifiers: LCCN 2016014264 | ISBN 9780226356068 (cloth : alk. paper) | ISBN 9780226356235
    (e-book)
Subjects: LCSH: Mood (Psychology) | Emotions—Psychological aspects. | Emotions—Social aspects.
Classification: LCC BF521 .C36 2016 | DDC 152.4—dc23 LC record available at https://lccn.loc.gov
    /2016014264

♾ This paper meets the requirements of ANSI/NISO Z39.48-1992 (Permanence of Paper).

For Mrs. Leach,
Ridge Avenue Elementary School,
who set the mood for arts and crafts

How it would interest me if this diary were ever to become a real diary: something in which I could see changes, trace moods developing; but then I should have to speak of the soul, & did I not banish the soul when I began? What happens is, as usual, that I'm going to write about the soul, & life breaks in. —Virginia Woolf

Almanack

perhaps (i) a variant (with specific semantic development) of classical Arabic *munāḵ* place where a camel kneels, station on a journey, halt at the end of a day's travel, hence (in extended use) place of residence;

The assumed semantic development from the concrete classical Arabic senses of the verbal noun to the sense 'calendar' has a parallel in the semantic development of climate *n.1*; in fact, *munāḵ, manāḵ* is the usual modern standard Arabic word for 'climate.'

Compare Occitan *almanac* (1548 as †*almanatz*), Catalan *almanac* (14th cent.), Spanish *almanaque* (first quarter of the 15th cent.), Portuguese *almanaque* (15th cent. as †*almenaque*), Italian *almanacco* (a1348 as †*almanaco* in sense 1, 1725 in sense 2); also Middle Dutch almanag (1426; Dutch almanak, †almanack), Middle Low German *almanak, almenak, almanach, almenach,* etc., German *Almanach* (early 15th cent.; < Middle Dutch).

It has also sometimes been suggested that post-classical Latin *almanac, almanach* is derived < Hellenistic Greek ἀλμενιχιακά (neuter plural), denoting an astrological treatise (4th cent. A.D.: Eusebius *De Praeparatio Evangelica* 3. 4, citing Porphyrius concerning the Egyptian belief in astrology, in horoscopes, and so-called lords of the ascendant, 'whose names are given in the *almenichiaka,* with their various powers to cure diseases, their risings and settings, and their presages of things future'). However, this Greek form in the text of Eusebius probably shows a scribal error for an original neuter plural noun σαλμεσχινιακά, which is of unknown origin.

Further etymology uncertain and disputed.

CONTENTS

*"A Mood Room Gallery" (a portfolio of color illustrations by
    Rosamond Purcell) appears after page 230*

on the street where you live

Every figure presupposes a background against which it appears as a figure—
this elementary truth is easily forgotten, for our attention is normally
attracted by the figure which emerges and not by the background
from which it detaches itself.—Didier Anzieu, *The Skin Ego*

Who can remember or even cares about the actual plot of the famous AMC serial *Mad Men*? I certainly don't. It's the *mood* that the show faithfully recounted and re-created, the interior and exterior palette that Matthew Weiner and his crew crafted—all of it barrel brewed like fine whiskey—that lingers, leaving us woozy in retrospect, dazzled for a spell but also haunted, much like the show's main character. This is what counts. This is art that gets under our skin.

For me, all that *Mad Men*'s seductive patterning required to get me hooked was the endnote struck by the very first episode: a philandering Don Draper looks in on his sleeping children. The camera is hardly given more than a moment to hover there, and with it, our own caught breath, when the tensely lambent scene cuts to the upswell of the 1956 hit single, Vic Damone's rendition of Lerner and Loewe's classic song from *My Fair Lady*, "On the Street Where You Live."

The show tune is all of one stanza long, but I don't know if there's a more successful lyric in an American popular lexicon for conveying that most particularly transformative of moods—the sick-making, suddenly swirling trance of the mood of being in love. It's a charmed lyric that gets falling in love right by picturing the lover as goofy and atilt and by relaying something fundamental about states of mind and the rooms in which we dwell; mood shifts and the atmosphere we breathe; surges of feeling and the ground that we blithely travel. Something has changed—irrevocably—and Lerner gets at that with a repetition, more poignant for being simple, of the little word "before." We'll recall the stirring opening lines: "I have often walked down the street before. But the pavement always stayed beneath my feet before." The first line sounds a lot like the second line except that something as unlikely to be noticed as a chink between cobblestones has unaccountably shifted. Now the lover sings of feeling all out of scale with his surroundings but also bizarrely in sync with them, tall as the buildings on the street where the beloved lives, certain it's the only part of town where lilacs bloom and larks unleash their song.

If the song works to stir a mood of recognition in me, that's not only because, like the next person, I have been in love and sought to remain in love. It's also because I'm intrigued by the way that we align the state of being in love with a mood of being alive even as to be in love is to be apart from ourselves, least like ourselves, if not temporarily insane. Do we have to be taken out of ourselves to experience our most heightened moods?

Now you might want to say that being in love is not exactly a mood but a calling and a blessing and a lightning-striking alteration that has the effect of beautifying us rather than mauling us, though it might do that, too, whereas moods are much more quotidian even if, like being in love, moods are states impossible to sustain. To be in love is more akin to being under an influence with suggestions of hypnosis requiring a Svengali-like object of desire. But what if I said that in order to write or make art one must be in love, not with an individual per se, but with life itself. That a particular faith must inhere, underscored by joy, even if what one writes about is hopelessness or human suffering. And what if I said that we are always under an influence—and preposition-less influences at that—though not always happily, and that such ever-present, vaguely perceived slow or fast-moving clouds constitute our ever-ephemeral, powerfully indicative moods?

There are so many ways in which Lerner and Loewe's "On the Street Where You Live" transports me. (Thank You, Matthew Weiner!) In my 1960s childhood, the street where we lived was broken. It was filled with people trying to make ends meet but without the means to do so. It was a street bordered by empty lots and rooms to rent: vacancies. It was a street on which the almighty dollar hung in the heavens in place of a sun or a moon: could anyone earn enough pennies to keep the rain out and the kids fed? Could one ever feel full—in love—or would a mood of lack, underscored by desperation, win out? It was a street where people made things—there was an obsessive welder of wrought-iron fences literally bent on reproducing the regal loops and curves of his native Palermo; there was a gardener (my father); there was a soprano that lived next door who practiced octaves with her voice now velvet, now satin, no matter the noise and the dust that crowded the street where we lived. In one of the street's row homes, my mother played and replayed the complete soundtrack of Lerner and Loewe's *My Fair Lady*, with which she also sang along.

If a mood can be created out of album covers, this was it: I can still see a cartoon rendering of a man's overly large and awkward hands as he manages the spindly limbs of a female puppet, frilly dressed and feathered with pursed lips—*My Fair Lady*. The album cover was greenish-gray and the golden strings rhymed in my mind with the crisscrossed geometry of a black fire escape imprinted on a red background of the album that was *West Side Story*. My mother belted out the lines from another musical, *South Pacific*, "I'm gonna wash that man right outta my hair!" while a bald and burly Yul Brynner held his ground on the cover of *The King and I* from within the precincts of a palace. During this period of my life, I was my mother's manager, on the lookout for the effects my mother's moods might have—on the rooms we all shared, on herself, and on the street where we lived. Would my mother survive her feelings? Would we?

My mother was neither a boxer nor a stage actress—my managerial skills weren't cut out for such—but a feeling-ful lover of life who could suffer crying jags as easily as unaccountable laughter. My mother's moods lovingly rocked me—especially in the form of her voice; my mother's moods revealed themselves like a peacock's spray of feathers from a far-off land where a bit of blue sky was suddenly delivered on a platinum platter ringed with emeralds. My mother's moods were reckless.

All of those romance-laden albums were better than Harlequins. They were more original, considering the poetry of their lyrics hitched to the sad or playful turns of their scores, and it wasn't a figure of unfulfilled longing or domesticated desire that my mother struck. A *real* love affair was afoot in our house, but it was a love affair with voice—could one match one's voice with the voice of Julie Andrews, Rita Moreno, or Mitzi Gaynor?—and it was a love affair with music whose effect was to make the street where we lived beam with some sort of bursting sunshine, transforming a cramped lane into a prairie of graduated feeling, of "corn as high as an elephant's eye," with "plenty of air and plenty of room," even if *Oklahoma* was not as regular a part of my mother's repertoire as "On the Street Where You Live."

In a Station at the Metro. Aboard the Titanic. In the Middle of Saint Mark's. Near Where the Lindens Grow. Home, Home on the Range. It's amazing to me how suffused with evocative power is the title of that song—*on the street where you live*—by dint of its emphasis on location, but also as a container, minus a subject: the beloved, you. Here's a song about the room being altered because a particular person is in it; the mood of an entire polis changed utterly by the presence of one inhabitant. But what's really poignant about the song is its source in solitude—which isn't to say it's just a song about the power of the imagination to incite a mood, though it is about that, too (let's face it: does the beloved here pictured even really exist?). What's moving is the scene of privacy the song invites us into. It's like watching someone dance alone in his rooms, and I begin to wonder if it's that unabashed, unembarrassed mood that fuels all musicals. The beauty here is that the singer is performing in public the mood that love has privately stirred in him—"People stop and stare. They don't bother me. For there's nowhere else on earth that I would rather be"—but he's performing for nobody but himself.

Remember "Singin' in the Rain"? That's the classical instance of a similar scenario: something about the surround has flipped, giving way to the sudden appearance of a cavalcade of uncommonly felt feelings, so much so that that great mood master—the weather— cannot toy with the singer's exuberance. Rain doesn't dampen his mood but is its perfectly cockamamie accompanist. The lover doesn't sing and dance *with* his beloved in the rain; he's once again alone, alone with a thought of you, alone with a mood that he dares, for our benefit and vicarious pleasure, publicly to display. Enter a movie theater, enter a TV screen, enter a record player, and my chronically depressed aunt Frances, one of my mother's four sisters, is momentarily happy: *Singin' in the Rain* was her favorite film, and before the age of the Internet, she sought it out, again and again. Of course the show tune's famous film scene in which a puddle-stomping and happily bedraggled, graceful and clownish Gene Kelly dances in the rain did not provide a cure for my aunt. Maybe that's why listening today to its lyrical cousin, "On the Street Where You Live," as I prepare to deliver these pages to a reader, out of reach, but there, makes me want to cry.

Maybe my reason for crying when met with such an exuberant song is that I'm one of those people who mistrusts a really good mood, those moments of pure unadulterated happiness that sweep in when we least expect them, and without the help of drugs, and that saddens me. And, no, the time-out-of-mind mood of well-being (show-tune moment: "everything's goin' my way!") isn't the effect of dopamine, serotonin, or the cookie that I ate for lunch; it's a jagged conjunction of a furred creature of a feeling and it cannot be explained. This is part of what makes it so all-consumingly right and wonderful: the extent to which it brings me to a place of self-forgetfulness and nonknowledge that is the mood, the whole mood, and nothing but the mood, so help me god. Consciousness has a way of creeping, and foxes too, but they're so much more beautiful for being sleek and eye-wise in their regard. Life breaks in to counter or question the mood, to read it as a part of a sequence or a larger frame, skewered by the arrows of a past, present, or future path or judgment or claim.

I think that's what brings a twinge of pain today to my hearing Vic Damone's sweet voice, but there's also a line that turns me tourniquet-wise that might have no effect on you. It's the lines that go: "Just to know somehow you are near. The overpowering feeling that any second you may suddenly appear." Call me a sad sack, but those bright lines bring to mind my mother who just turned eighty and in whose presence I learned to love and fall in love because wasn't that what was happening in the mood space she'd create with all those cinematic soundtracks? I'd watched her being in love and studied her along with those record albums, certain to eke out a drum roll that might ply the right color or hue capable of changing night into day. My mother has just turned eighty, but I am many, many miles away. Surely the hopefulness of a sudden appearance, and the mood of overwhelming anticipation that comes along with it, are fueled by the reality of disappearance: don't ask me to bear life on this planet without my mother in it. And, Jean, my partner-lover for over thirty years, she's "near" in the "suddenly appear" but that doesn't mean our closeness isn't occasionally broken by a chilly feeling—a wind-swept mood that is just the obverse of a wind in a sail or the gliding gladness of a bicycle breeze. It's a feeling that surprises me sometimes usually around bedtime, or at least amid the abiding mood of warm sheets, that one of us will someday predecease the other, one will be required to wander around inside the hole of that and find a way to climb out with a new friend called absence as company.

Such responses and the moods that spawn them are personal and quirky and as singular as Vic Damone because, as it turns out, his distinctly powerful quiver of a voice, his heartbeat-raising version of this song has nothing to do with the street whose beloved inhabitant the song pays homage to. A convert to the Bahá'í faith early in his

classically troubled career, he has said that the fervor with which
he sang those particular lines—"For, oh, the towering feeling just
to know somehow you are near, the overpowering feeling that any
second you may appear"—emanates from his conviction that he's
singing to his god. In the rendition of the song that *Mad Men* calls
upon, as though to underline this point, he sings those lines unconventionally as a prelude to the main verse as well as in the middle of
the verse. What gets Vic Damone going is his god, but what gets me
going is Vic Damone.

It's not just the song but the singer that effects the mood I'm
after. For me, it can't be Peggy Lee's bossa nova version complete
with bongos, trombones, and toreadors sweating under their oddly
shaped hats, dressed, as the saying goes, "to kill." Eddie Fisher might
strike a sweet tenor of a version, but all believability is lost the moment he pronounces "lilacs" like he's a baby lapping "loll-i-pops."
Poor Doris Day gives us the song as a Shirley Temple in adult drag.
Willie Nelson's lisp is endearing, but his preoccupation with his
voice overrides attention to the lyrics. Mario Lanza confuses the understatement and vulnerability of the song with the high tragedy of
Pagliacci: a mood of sympathy is all that I can muster in the hope
that someone will unstick him from the cave of his throat. With
Dean Martin, it's hard not to picture Jerry Lewis "suddenly appearing" like a bucktoothed hand puppet, and with Placido Domingo,
I give up the minute he mispronounces the crucial line, however
sincerely he ululates, "on the street where you *leave*."

Only Vic Damone gives us the mood of longing born of breath
and pause, from the part hush, part sigh of his only nearly arriving
at the opening lines, "I." . . . "have" . . . "often . . . ," to the free fall
inside the gap that enjambs, "all at once am I" (long pause) with
"several stories *high*," to the bleeding in of "by" with "I" in "let the
time go by" where he fails to breathe altogether, letting the note exceed his voice as container. Damone gathers a range of tonalities
into his mood net, and maybe there *is* something of the holy and ecstatic that he mines from within the kitsch because you might have
to call upon some power beyond yourself to make the leap available

as he does, from his own operatic surge—the sort of note that is ripped from one's entrails—to the shy whisper that makes me want to weep.

Will this song always have this effect on me or does it depend on the mood I am in? Can I create a change of mood in you from worse to better in the reading of this book? Is the best way to do that by reading to you—which is to say, by way of writing—or by splashing through puddles like a maniac, singin' in the rain? Would Eva Marie Saint have been less trusting of Cary Grant if Hitchcock hadn't directed that he whistle "Singin' in the Rain" in the shower from which he spied in *North by Northwest*?

The famous French philosopher Gilles Deleuze proposed in an interview that we can learn a great deal about the zones of thought and feeling that bound us by asking when and where we sing to ourselves. For Deleuze, it was when moving about in his rooms; when afraid; and when leaving. My mother, a poet, once described how her rooms oriented her: in her bedroom walking around the bed tucking in sheets, human verities occur to her. In the kitchen's narrow space, whether cooking or washing dishes, problems are solved, decisions reached. By rivers, creeks, and roaring brooks, she writes. In doctor's waiting rooms, she observes and notes. But there's nowhere in the world where she—and we—aren't always singing, and in large part, singing to ourselves. If Hollywood musicals treat singing as a break in our otherwise flatlined routine, that's only an effect of their bringing song out from the background where it always keeps up its beat and into the light of a different day.

~~~~~~~~~

What's more rife with mood? Television or the cinema? Janet Paige is dressed in a contoured green satin dress that clashes viscerally with her orange-red hair (think *Mad Men*); Fred Astaire is perfectly put together in a gray flannel suit and silver tie, offset by white-rimmed, brown pointed shoes. Together they pounce and prance rather than dance, leaping over an upright piano and crawling along a floor in the mad antics of a hilarious number called "Stereophonic Sound" for the 1957 musical, *Silk Stockings*. In this purposely graceless routine, Paige and Astaire mock the eye-popping colors and wider and wider panoramic screens of their day—the technological monumentality of things like seventy-millimeter film and the need for special projectors meant to entice people away from their diminutive TV sets and out into the surround sound of Hollywood musicals. In a send-up of the 1950s turn to Panavision, "or Cinerama, or Vista Vision, or Superscope," the singer/dancers have Cole Porter to thank for the cannily pointed lyrics: "the customers don't like to see the groom embrace the bride, unless her lips are scarlet and her mouth is five feet wide. You've got to have glorious Technicolor, breathtaking Cinemascope, and stereophonic sound." Seated at either end of a very long conference table, Paige and Astaire can't stretch their arms far enough to reach each other's hands, and the scale of the new technology forces them to slide on their bellies like flipperless seals to reach one another. At the end, the mock lovers swing from a chandelier, as if to ask, "Is this, Viewing Public, the only thing that will really bring you out from within your television-lit living rooms? A more and more extreme calisthenics of love in place of an atmosphere of romance?"

There's no question that the rooms we create to mimic our moods are undergoing a sea change. Now that viewing audiences are pressed deeper and deeper indoors, and in spite of the allure of IMAX where we can plummet together headlong into volcanoes, it's fascinating to see certain contemporary filmmakers return to seventy-millimeter film almost as a hysterical response to a mood fast vanishing: the sorts of moods we used to know by way of the movies where we sat side by side with strangers in the dark. "Movies," for all we know, might soon be replaced with "light shows"—my most recent experience of which required neither a four-walled theater nor a narrative, but took place amid the elements where we wandered in the rain, no less (there were buskers singing there), and where a city's monuments became the projective surfaces for wondrous forms, only visible at night. Just like in a movie theater, a building or booth hides the projectors and the digital sources of such extravagances, but sometimes the light bisects other things in the landscape as when, for example, the white gush of a fountain morphs into a square of purple plains. "The Festival of Lights" I experienced one fall evening in Berlin might be a replacement for the candles we used to light to cope with the effects on our spirits of a darkening season. Now the imperative is to venture out to join a group of fellow wanderers inside a holograph, for that's what it felt like to me, all of us, including the natives, all a little more lost than usual inside the illusion—oh, but the city had already met me that way, as though most of its buildings were made of cardboard, as if I'd been residing for a spell not in real urban spaces at all, but in one big Hollywood set.

Consider a relation between moods and rooms as reciprocal: we experience moods as containers of ourselves and we create rooms *in their image* at the same time that we create rooms to *alter* our sense of those invisible containers: our moods. Rooms are changing, as are walls, since the advent of the computer screen, but we may not really understand what is meant by that until the screen is done away with altogether and walls become the reflecting surface for a much less cumbersome device. When public and private surrounds—but who knows how those will be parsed?—become filled with the data streams of our own and our fellow men's digital worlds, when self-management systems that pass as forms of sociality come to constitute our environing surfaces like so much holographic wallpaper, it's anybody's guess what will happen to mood, both personal and shared.

I experience the problem acutely in train stations. I experience it as a surcease of waiting. It's not just that I stop noticing where I am, especially in a place I've never been before; it's that, since the acquisition of my iPhone, I wait differently. To wait: it's a verb we use to draw a line between child and adult. Waiting: it's a threshold space, a neither-nor locale where nothing happens and everything occurs: a place that houses the imagination and that makes possible art. I remember well the feeling of strangeness—the unfamiliarity of my own coat cuff or shoelace especially in train stations en route to the unknown. I remember the feeling of making my way, and settling in, coupled with unease; of maybe talking; of sitting silently and looking.

So what if Philadelphia's Thirtieth Street Station is an art deco masterpiece. I could be anywhere, because I never really have to leave the world I carry in my handheld device. It's my pacifier, and I am as coddled by the known world that I access therein as I am entranced by it. My iPhone and the content I can access through it is a seductive placater that never allows for orgasm but produces a great deal of dithering in space and a plenum of forgetfulness.

In what sense are iPhones replacements for our moods as holding environments? I'm not sure. I only know that there is a kinship between the ability for a person to be open to strangeness and the ability of a person to be able to wait. Our moods are strange to us yet abiding. I want, if nothing else, to produce a form of strange beauty in mood's name.

~~~~~~~~~

Is it possible to give mood a sound form and a palpability without pinning it down?

An Anglo-Saxonist whom I know tells me that the word "mood" comes from Old English "mod," which has been widely written on since it is rather indistinct. It is both mind and heart, but it might even refer to a *place* in the head or the breast. There are lots of mind/heart/thought/coffer/chamber terms in Old English, she explains, where "mod" or "sefan" resides in the "breost." So, moody used to mean passionate, "modig," brave, courageous; it was different from "hycg," "thought," in engaging the emotions. It was part of many compounds such as "modecearig," "mood-care-ish: "sad," "mournful"; and "ofermod," meaning "proud," "arrogant." Mood seems linked in the language's earliest days with wrath or courage, conviction or passionate grief. I imagine it not exactly taking up residence in what is thought of as the ancient or animal part of our brain, but a sound form as old as the idea of fermented honey, coincident with the concoction of gods with names like Thor, before the days of razors when people were prone to bushiness and drank from goblets shaped like helmets.

"Mood," I think, is a word as old as the stirrings—bolts of pain, rumbles of glory, gut joy of survival—that lathered the hearts of nomadic tribes. To take it on might require opening a door with no hinges, made of nothing but originary rubble and the dust whence we came, dead particulates that might have once been tools—that's imaginable—but how about the granules of what were "feelings"?

In time, the word "mood" came to refer to a prevailing *or* temporary state of mind, and, depending on the model of psyche or soma that marked the age, it referred to a humor, temper, disposition or, my favorite definition, "air." Not until the early twentieth century does mood come to apply to a group of people, a collective body, or a pervading spirit or atmosphere.

In the twenty-first century, if I tell someone I am writing a book on mood, the assumption is that I am writing a book about depression and that I must be, by extension, depressed. Maybe it's hard for people to imagine writers being interested in anything other than themselves, and that's what fosters the second equation. In a book about awkwardness, I addressed everything but its more commonplace kinship with embarrassment and shame. With mood, I'll no doubt press depression's pedal on mood's vast pipe organ, but it's not depression that takes me to mood. I have a hunch that we may be entering a moodless age, but that's not the same as a depressive, depressing, or depressed one.

In the early twentieth century, "depression" entered our lexicon as a psychiatric disorder, though some more interesting terms for potentially similar states predated it by centuries, like the sexual-sounding "descence"; the fluttered frill of a "tristesse"; the spiritually imbued "soul-sickness"; the exactingly physical, "jaw fall"; or the bluntly negating, "unlust." "Mood" spelled backwards is "doom," and unaccountable fits of gloom or bad temper to which the word "mood" applied enjoyed a range of, in many senses, playful-sounding epithets before the word "depression" descended to seal the matter, as though lowering a lid on a tomb of human consciousness. In the sixteenth century, words like "the mubblefubbles," "mulligrubs," or "mumps," the "sullens," or "momurdots," in the eighteenth century, the "mournfuls," "mopes," and the "grumbles," in the nineteenth century, the "doldrums" have a touch of mockery about them—even an onomatopoetic baby talk—that, once applied, might enjoin the person so described not to take himself—or his moods—too seriously, as if to say, the social body needs you: these words, in a sense, are relational. Angry or irritable moods are equally wonderful in their wordy wordlessness, with signifiers that seem more like mere phonemes relying mostly on the flick of a tongue against teeth as in "tetch," "tantrum," "frump," "strunt," "tiff," "tift," "tig," "tout," "snit," "pet" (cf. "petulance"), and "miff." In almost all cases, words like "mulligrubs" or "mubblefubbles" required a definite article—"the"—or a possessive and had in mind a state a person was *in*: she was in her "mulligrubs."

If these words seem more particularizing of mood, the word "depression," a more global descriptor, seems depersonalizing. On the contrary, the dropping of articles—we'd never say, he is suffering from *the* depression—inaugurates a type of person; with "depression," we enter the age of personality, a conception in toto, disseverable from the self and therefore more likely to give rise to one. Oddly, depression is both more individuating of persons and less exacting a term. It's as though the price of becoming a modern subject was to sacrifice variegated feeling and grain. To be sure, our language for moody states of mind has become less precise at the same time that we speak more certainly of moods as something pharmaceuticals can treat. Mood disordered, we live in an age of mood-altering, -stabilizing, -elevating drugs: in short, we seem to think we can regulate moods, or that moods need regulating in the first place. Benightedly bandying acronyms, like brands and flavors of personality and minds, we presume authority without caring to understand, our discourse of self-knowledge shrunk to the size of a pill, as easy to swallow as a self-help manual.

I could say I want to find a language for mood because "there will

be no greening of the economy, no redistribution of wealth, no en-
forcement or extension of rights without human dispositions, moods,
and cultural ensembles hospitable to these effects." Because, in other
words, political change can't happen without psychic change, and
that's much harder to effect. Because there are companies whose
work it is to make stores smell a certain way in order to inspire consumer confidence—to put you in a buying mood. Because data show
that people are in a worse mood after they have been on Facebook
than before because they feel they have been wasting their time, and
most of the people I know are on Facebook 24-7. Because I know a
woman who says yogurt can definitively improve your mood but the
pharmaceutical companies won't hear of it. Because in the twenty-
first century, groups have sprung up that call themselves whispering
communities who don't exactly gather but sit alone before their com-
puters in order to watch or listen to videos in which a young woman
crinkles the pages of a book across a very long period of time or taps
her fingers on a tabletop while whispering banalities into the screen
with the intention of creating an ASMR (autonomous sensory merid-
ian response a.k.a. an "unnamed feeling") in her audience—a deeply
relaxing tingling sensation that also presumes to elevate one's mood.
Because I know a guy who wants to promote weather modification as
an instrument of war, and I want to understand where the line breaks
down between allowing for the destruction of our environment and
legislating against purposely tampering with the weather. Because a
literary critic out of Berkeley has coined the phrase "ambient literature" to describe a form of contemporary Japanese writing meant to
create a calming effect on its readers as antidote to national stresses.
Because mood has something to do with gravity—such interdepen-
dent force fields, moods "lift," they "swing." Because I hear that physicists are beginning to theorize consciousness as a form of liquid,
solid, or gas. Because to exist is to be light and to die is to weigh—or
do you imagine it the other way around? Because I imagine people
who suffer from depression reckoning daily with the weight of Being
(as the condition) toward which life moves, admitting the fact of the
matter we'd prefer to avoid: that lifeless things and newborn things

fall and that some other must be called upon to carry us, and not necessarily a lover, but that it's always a stranger who lifts and lowers our
heft in the beginning. And in the end.

~~~~~~~~~~

Think of the diminished range of tones that we're allowed in the
course of a day or the course of a life.

When was the last time you treated yourself to an excessive
tone or an extraordinary one? Or the last time you were aware of
a shift that could unleash a liberating feeling? Shouldn't our towns
and countries—the streets where we live—lend us an infinite tonal
range rather than the cribbed playing cards of happy, sad, angry, and
bored? Where's the guy responsible for naming paint colors when
you need him? The wall colors available at our local hardware stores
get to be more venturesome and subtle than our moods.

I derived my first inklings of the need for a book on mood on a
drizzly day hidden inside a library with a book whose central image
moved me entirely: it was the idea developed by French psychoanalyst Didier Anzieu of what he calls the "sonorous envelope": the protective and precarious, requisite and porous, holding environment
(literally, a skin) that is lent us not by touch alone but by the *voice* of
our earliest caretakers. I love the idea of our sense of embodiment
being established by the sound of another person's voice, and I can't
imagine a telling of mood that doesn't try to account for sound and
the way that sound environs us. The real question isn't whether you
prefer the operating systems of a MAC or PC, but which sound you
prefer for how it sets the tone of your day with each depression of
the "on" button. Tone is everything, and I write to create an alternative to the impoverished mandate of our daily tonic repertoires.

If I could sum up the aesthetic challenge of this book, it would be
this: not to chase mood, track it, or pin it down: neither to explain
nor define mood—but to notice it—often enough, to listen for it—
and do something *like it* without killing it in the process.

I like to call what I'm doing here "cloud-writing," but it may as well be a form of essay writing—that nongenre that allows for untoward movement, apposition, and assemblage, that is one part conundrum, one part accident, and that fosters a taste for discontinuity. Cloud-writing and essaying meet at points of attention and drift, inviting us into the precincts of immersive absorptive planes, inciting altered states, sidling up to a reader and intimating, reluctant to explain.

The halt at the end of a day's travel. The point at which the camel kneels to let you disembark. I was as surprised as you are to discover this etymology of the word "almanack." Some surmise it is of pseudo-Arabic origin, a word invented by medieval cosmologists to lend an authoritative stamp to their books. Others want to hear in it the Spanish word, "alma," for "soul." From Ben Franklin to the *Farmer's Almanac* of today, the almanack is a revelatory book and a book of secrets. A book whose tidings we look out for and consult from time to time. A book that glimpses movements—of stars and of puddings. That predicts days, or times of day, to be avoided. It's a kind of compendium whose principles seem to be calculability (calendars, timetables), whose potential is to help our knowing when to plant and when to harvest but that is undercut by some incalculable flavor of things. It's a miscellany built inside a chart-bound form. And we like the unpredictability of that. Of a voice speaking to us. A book that works as a companion, one eye to the earth, one eye to the heavens. A book to wander in a desert with. A book that perpetually dismounts and begins again. A book whose only requirement is that we float into and out from the streets where we live, pausing long enough to feel the mood beneath us shift.

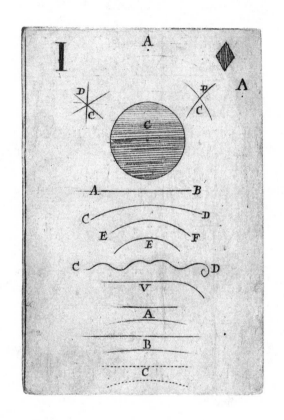

(*elements*)

Atmospheric changes, provoking other changes in the inner man, awaken forgotten selves. —Marcel Proust, *In Search of Lost Time*

mood of perfect kinship

Where did this convention originate? The habit of photographing a child—never children in the plural but just a child solo, as if she were suited prodigiously to perform an aria—in front of a bush or shrub, a spray or display of flowers? I remember a friend's home movies and our viewing them with her, decades beyond their making, on the edge of our seats with expectancy, and she, just a little terrified because she didn't know what they might contain though she knew they documented a period after her mother had left her father and had subsequently left her for long periods with a great-uncle and -aunt. The couple, who were childless, doted on her and were, at the same time, very proud of their garden: the impulse to document the two as interspersed frontispiece is unrelenting, and it is poignant to watch the girl attempt to resist her placement, to try to leave the magic circle inducted by her being filmed, never when she's playing—and she clearly wants to play—but ever posed before the flowers. Here is the girl in pink chiffon squinting afore red azaleas; the girl in an orange clown collar pointing to red tulips rimmed with black; the girl in outlandish green pantaloons (the aunt had made them for her) posed sullenly before the bayberry bushes and holly.

My own garden enjoys an array of statuary placed pleasingly to surprise the pace of its undulating brights and shades—such small figurines of a Roman Diana, a winged gargoyle bent over a book, a Poe bobblehead (unsuitable for indoor display) do their best work if somewhat hidden to create the mood of a hushed, separate life in-side the garden's manufactured paths: they are bits of culture hidden inside of nature but poised as though they were birthed there.

When the adult photographer places a girl as if she were a piece of statuary in the garden, does he want to show off the garden by way of the girl, or vice versa? Or is the idea that the child, a flower herself, is the figure best suited to the company of flowers, in which case the convention is a form of "family photography" par excel-lence? If only I could add this live and perfect flower to the garden, the photographer seems to say, I could create a mood of perfect kinship.

"Go stand in front of the flowers," the adult commands the child in any number of private or public gardens across the land, even though the father's collar be rimmed with sweat and the day made hotter by the lit tip of his cigarette. Usually impatient, he maintains a nimble hold on both cigarette and camera; he finds a way to steady the viewfinder and shoot.

The child, meanwhile, isn't sure what she's doing there or what she's supposed to do alone before the flowers, so she quells her de-sire to fiddle with her hands, only noticing many years later the un-canny correspondences of the formal composition: her head is ex-actly the same dimension and size as the hydrangea pom-poms, and just as heavy on its stem.

Now let us admit that flowers cultivated or wild, subtly scented or loud, sentimentally imbued or inconclusively ebullient form an atmosphere, and that this atmosphere is a mood, and that for ev-ery photograph of a child—usually a girl but occasionally a boy—so posed, there is the question of what mood the child was in when she was asked to merge with the petals or accompany them.

"I just want you to be happy," each photograph of a child before a bunch of jonquils might announce, no matter how depressed the photographer, how desperate, or how mad. Herein lies the beauty of such photos, however stilted they may be, because eked out or etched in, the child's unchartable mood is somehow apparent, even coaxed to life, by her being set deliberately inside an atmosphere of bushes and blooms. The child's *distraction* is her mood, and it cannot have a name so bland as "happy."

Are we born having moods? People seem to think so, and that such moods stick for a lifetime: "You were born laughing!" "You were born crying!" Emotive emissions become confused with moods, and the attribution of a good or bad mood commencing at infancy must be very hard to shake. The truth of the matter is that we don't come to have a mood and aren't capable of being in a mood until we are photographed in front of Aunt Betsy's prize amaryllis or the ten-drilled fairy world of Uncle Septimus's colorful and high-climbing sweet peas; the sullen locution of some shady ground cover or the occasion of one suddenly upright stalk; or a bumble of snapdragons cultivated in public conservatories, bird-less and bee-less, beneath snow-covered glass.

I'm not yet three years old in the photograph my mother took of a hat that sits atop my head like a dome; a pair of black polyester suede Mary Jane's; and a dress embossed with orange wedges I thought that I could eat (fig. 1). It's obvious I've been sent to find myself in front of a many-tiered spread of mostly-white-with-touches-of-purple irises and roses, columbine and carnations in the Blessed Virgin Mary elementary schoolyard though I'm not yet old enough to go to school. By the looks of it, it's a date that will become a favorite to my mind for the pleasure I will come to take in being part of slow-moving processions, underlit by the scent of incense, and the strewing of daisies, counterpoint to the otherwise cold frames of my days in Catholic school. It must be May Day—that explains the statuette of Mary who hovers on a higher plane behind my head, and the attitude of the older kids: a girl in the background opens a schoolroom door into the yard as if to fly (May Day always signaled that the end of school was not far off).

What endures for me in the photo that I could not have seen as a child are the parti-colored corner vases that frame the arrangement but break the mold. What's vivid to me is that I'm not yet uniformed; I'm not yet indoctrinated. Nor am I innocent. The scent of the flowers must have been overwhelming—why, otherwise, have I shut my

(1. *"We just want you to be happy"*)

eyes and let my hands fall limp? On closer look, I see that my hands aren't limp at all but feel with pleasure the texture of my dress. I'm posed before the flowers in the photograph, placed inside an atmosphere, and clearly living in a space apart from it.

Who knows what the ground could mean to those tiny feet beneath those down-turned hands and the tangle of those spiraling curls.

I have to thank my mother for the generosity of her gaze—I know it was my mother who took this picture even though it was my father who manned the camera throughout our childhood years. I need to thank my mother for not commanding me, "say cheese," and in this way allowing no more than a hint of a smile to billow from within rather than be canceled by a broad-faced grin.

Certain photographs recur for us; like moods, they appear in the mind and then fade. How often this scene of being in the school-yard, but with the pleasure of not yet being in school, appears to me, and my need thereafter to find the photo again. As if to say moods are the faces we make out of purposely closing our eyes, while all around us the flowers are in bloom. This is the place where mood, for me, begins. My Moodday apart from my birthday.

mood modulations

Moods, in the strong, Schumannian sense: a broken series of contradictory
impulses: waves of anxiety, imaginations of the worst, and unseasonable
euphorias. This morning, at the core of Worry, a crystal of happiness: the
weather (very fine, very light and dry), the music (Haydn), coffee, a cigar,
a good pen, the household noises (the human subject as caprice: such
discontinuity alarms, exhausts). —Roland Barthes, "Deliberation"

&

Once I removed, set to one side, ignored, reduced the patina of my
repetitions, I began to have feelings, but not necessarily moods.
Bug, my cat, definitely has moods—sometimes she eats a lot; other
times, goes on a diet. She pads, or plays, or sleeps, and for her, the
day seems to have its own seasons. I wouldn't know a mood if it hit
me on the head, or only if it's a bad mood, or irritable mood. "Mood"
proposes a break in the chain that comprises personality—if I'm in
a "good mood," this assumes this is not a regular state for me. Suddenly, I take flight, have wings: "*You're* in a good mood!"—we've all
heard this, as though, again, a pattern has been broken, we've taken
more than our share of pie, we've refused to stay in our designated
place, we have dared, *how dare we*, we've been indignant. When I
say, "I don't feel like it," I really mean, "I'm not in the mood," and it's
anybody's guess if I ever will be. But look: the sky has changed again
without me in it while I was looking at my book, at my page, or turning inward. Even as the clouds—that outward mantle, that reservoir

of fluff made manifest, or their departure, will continue to affect, in fact, determine, my mood.

☁

    Mood triplets (recipes)
    The smell of dill,
    A burgundy carpet (stained),
    A hibiscus blooming indoors (a plant with one bloom)

☁

Life as a series of moods—like Picasso's Blue Period—tonal registers of being. Or a series of rabbit holes, or doors within doors, moods as rooms, rooms I have known in order to arrive at the place where I hear things.

☁

    Moods are cubbies, and we are their cubs.

☁

My four-year-old niece Sophie whispers in my ear: "I want to be an RCMP officer when I grow up."
"I ate a piece of candy from the floor."
"I have two Ariels."
"I hurt myself (boo boo)."
Versus the stage whisper, cries and whispers, wisps. Whispering woods, or pines or, auditory hallucinations.

☁

The droplets of rain dripping from a geranium leaf are perfectly out of sync with the note made by way of intervallic drop on aluminum, skittering, bouncing upward, soundless, or outside my range?

☁

We are always in a mood of one sort or another. We are never *not* in a mood, but certain conditions need to inhere for a mood to be created, and the same is true for being taken from one mood to another. For example, this morning with the blinds drawn, the fact of radio music apparent in another room—never the room one is in—but adjacent, nearby—could, if it were left to play, determine the mood for years to come (or so it seems), not so much because of the fact of a violin but because it sounds like it is being played alone in a barn while another set of hands, invisible, with blind conviction, searches for water. The mechanical thud of a key striking a string opens a lock the way the pipe resists the contour of the cloud in a painting by Magritte—the pipe is the cloud, the sky, and less like a belt buckle opening or closing to start a day: prongs in holes, tines in slabs of egg whites, opiate receptors. Each plunk is the water sought and the search for it—the quester's plunge. The sound resists the image you want for it or to ascribe to it—of prongs, of tines—instead insisting on a black cardboard stencil silhouette of sky.

☁

Moods are our ontological compartments (shelves for selves), otherwise this A.M.'s mood idea lost at breakfast table with cereal and supplements.

☁

It's possible that moods are not fluid even though they exist on an ephemeral plane. We expect them to be mesh-like when the fact of a mood is its being in place, a place, immovable (but not unchangeable).

☁

The verbs that attend "moods"—to lift. To pass. Or be fleeting—it was just a fleeting mood. To bring you down.

☁

The mood of the way Jean mounts a bike (more like a horse—two feet on one pedal!). My approach: more like jumping onto an already-in-motion carousel. Could be part of a mood query.

☁

Academic moods (yik).

☁

Three different sounds from which a mood-cloud could be built: "scraping."

1. from bed, as heard in A.M.: Jean is sweeping a rug in cabin
2. again, from bed, making popcorn in deep pan on electric stove in cabin
3. Bug using litter box in cabin (swipes)

☁

The mood of food, but in a way that undoes or shakes up "food" writing. In Günter Grass's *The Tin Drum*, a character "who turns feelings into soups."

☁

If a leaf falls in a rappelling-like fashion, accepting and repelling its swift descent, or if it succeeds in a succession of flips, is this somersault, edge over edge, enveloping, a mood?

☁

Moods are borne of methods: washing dishes, with or without singing; a server turning the teacup handle toward or away from the person being served.

☁

Highly privatized imagery: two shuttlecocks fall from the "branch" that they protrude from, clown noses hitting the floor, fake bird bestiary—*thunk*—like or unlike the sound of a bird hitting glass?

☁

Pie à la mode . . . or à la mood? Where there is food, mood cannot be far behind.

☁

A world made of clouds isn't formless—there are ridges and peaks, incommensurate intertwinings, appositions, and breakings through. Even if there is not density, there is form: a mood is like this: it's a form without density.

☁

What does it mean ontologically for me and my cat to be flying in an airplane together above the clouds, looking down? The question it comes down to: What is a mood's relationship to *matter*?

☁

Sexually speaking, I have ever suffered from being eminently seducible. I can always be made to be "in the mood."

☁

Mood lists
Caponata
Dessert
Mirto—deep plumb
Rosolio—light pink
Zibibbo—raisined
Bread
Mood of hubbub
Don't forget
Tea
Teaspoon
Essay
Jim's letters
Cards
Stephen's address
(Hilarious) an idea for an essay—what was it? Must remember it
LA–Cleveland (snacks)
Cleveland–Providence

☁

Mood: Tint

☁

Writing: a call in response to something greater than oneself (external) and yet that issues from some part of the self otherwise cut off from one (there's the rub) and that knows how to honor the daily in lieu of the grand. Elemental. Watercolor.

☁

1. Clouds and mountains, mountains and clouds, seeing without seeing: deep in a dram of you: locked in by clouds.
2. Bug's eyes looking up as I look down from inside clouds below us!
3. What makes it necessary to make polenta and mushrooms now, here.

☁

With today's cloud cover (cloud clover) we are truly underwater! And I could if I wished poke my nose (dolphin) into a different atmosphere in search of a different air.

☁

Today's clouds: a vast, fast-moving sheet made of cobbles.

☁

If the mood of a day is a tone—the tone of my days—then we must in some way be talking about a day's (diurnal) sound. The day's tonic.

☁

Mood: cloud cover.
Mood: a room with no walls.

Is the sky what is up above, or all around us? And if the latter, does the idea liberate or threaten us, embrace us or hem us in?

In the family of origin mood space, did I ever—and if not, why not?—listen to voices by putting my ear to the table, or the wall, or the floor? To my heart?

Some moods are the residue of a never completed cry—a wail?

Moods are contagious, or confined. Mass hysteria, e.g., or depression. Laugh and the world laughs with you; cry and you cry alone.

If moods are only, by definition, prevailing states.

The way my mother moves among tonalities in a letter (see recent 8/15) until a letter becomes a song: is this the range I've been trying for in my writing all these years? An aesthetic?

Falling to sleep, steeped in memories of kindergarten jumper (the softness of a particular yellow sweater and its lone pearly button beneath the stiff wool plaid)—the photographer tamping down my curls (why did he brush my hair?)—there was a black velvet ribbon in it (have all of those cells, those selves, been replaced so that I can no longer claim to have been that girl?). The memory coincides with a very different memory of waking up bald during chemo. Such indignity. Where do these coincide, these particular instantiations of being? Where one was trying to be sensate while something was being done to one's hair? With it, vivid memory of Mrs. Leach, of girl named Yvonne, who repeatedly picked her nose and smeared it on desk. Ridge Avenue Elementary School. The teacher's house! I am a special student. And how it felt at six to make that visit to Mrs. Leach's house with my mother. Three muses. Lawn Furniture. Lunch. Is to be awash in memory at bedtime to be summoned by or submerged by mood?

☁

Is there anything a mood *can't* be applied to? Think of the mood of a fever. Of solitary play (girl with scooter on asphalt); of roses, red or white; of this armchair; of anything animate or inanimate; of screens, rough winds, a dress, a bow, leaf veins, roof tiles, hammering, cat fur, telephonics, of opals as birthstone, the ice cream truck's jingle, a coo or call, an umbrella to block the sun, the mood of a sail, of wind and mist, of pie-eating contests' stomach grumbles.

☁

Why is it so difficult to reestablish a mood once it is "broken"? And is this most often the case in sex and love?

☁

The same thing that water does to sand, air currents are doing to the sky today. The calibrated ripples of a washboard abdomen. The sky as skin and therefore tympanum.

☁

On a day of confiding they were "out of sorts," person A said, "I have competing emotions and conflicting desires," while person B said, "What I feel is that I'm surrounded by an amorphous cloud of dissimilar particles and not feeling so great."

☁

In English, we say sunny mood but never moony mood.

All it takes is *x* to put me in a *y* mood: nearly burning myself on a cheap lamp in order to turn it off; cleaning someone else's hair out of a bathtub drain.

It's not my fault that mood allows for hedges and shade but isn't good at hems or borders. That it's tied up with the very place that wind comes from. If we can't chase mood, we might have to smoke it. To roll up these pages or neatly lick each end page into something you can inhale.

Dusk gets to be the time of day we most associate with mood, but that's only because we're most likely to notice transit points—to wake is to transition—while we drowse inside the calibrated panoply of moment-to-moment moods throughout a day.

I'd venture the day before a funeral to be more redolent of mood than the day of, and the day after, too, strikes a mood like no other day in which we've lived. Mood must be more at home in states of anticipation and states of blank than in the place where something happens.

We speak of an illness needing to "take its course," by which we mean that it is both out of our control and sure to pass; we, however reluctantly, allow the illness its own movement, and rest assured that it will arrive at a statistically predictable petering out. But mood doesn't work this way. Moods don't follow a chartable path or career. We can't grant them seven stages or twelve steps. A fever "comes on"; an illness is something we "contract." A mood, in contrast, "comes over us," like a cloud or a mantel thrown over the soul. "What's come over you?" we ask when we mean what mood are you submitting to, to the detriment of yourself and others? What change of mood or alteration is possessing you? On a brighter note, what are you allowing yourself to become?

☁

If moods are said to "lift," it might be because we ally them with spirits, vapors, genies, aromas, and invisible sorcery. Mood seeps in and encroaches, pervades and is dispersed, hovers and settles in.

☁

If moods are the invisible worlds we live in, we might have to trick mood the way that scientists did with air eventually to be able to visualize and weigh it, describe and manipulate it, believe in it and honor it.

☁

Pain obliterates the possibility of mood, making me think that mood requires a body in abeyance, an unself-conscious body?

☁

Ethnic stereotypes and mood: x are a cheery people, a gloomy people, a simple people, a moody people. "Melancholia is not French."

☁

The "debonair" is a person literally of good or decent air born. But what of you and me? We can put on moods, like airs, or breathe the mood air in.

☁

There is no such thing as a film that does not create a mood, but some films and filmmakers are known as mood makers.

☁

Is there such a thing as a manufactured over and against a "natural" mood? To manufacture a mood is not to still something fleeting but to create multiple forms of egress from the still.

෨

How often do we hear ourselves saying: "Depending on my mood, I. . . ."; or, "It all depends on my mood."

෨

To each person, the book she needs to read to get into a mood, the music he needs to listen to—today, Prokofiev and Henry James, tomorrow, not much different except for replacing Prokofiev with Rachmaninoff. The Russians! The Russians! They always reach me where it counts. It can sometimes feel like a soul massage of dark-bright thrums on black-white keys and struck hammers, of escalating weathers and cold temperatures in tempos of leaps and bounds.

෨

Julia Kristeva opens *Black Sun*, her book on depression, with Psalm 42, "Why so downcast, my soul, why do you sigh within me?" Kristeva sees the depressive as denying primal loss—identifying with it instead. Does the depressive fetishize mood? Depression is not so much a mood as it is a refusal to give oneself over to mood's unpredictable changeability.

෨

Mood as soup or stew—steaming?
Mood: a porous container, not exactly a sieve.
More like a web: beings capable of producing such: complex and beautiful forms meant to lure and trap prey.

෨

The ultimate mood music: an online radio station that plays nothing but piano practicing.

෨

What the clouds/mist are doing to the mountains today: you can't make desserts like this, though you try: layers of mousse, cappuccino, or soué. Grades and shadows (not exactly), caps: veins. With their tops missing, there, not-there.

☁

A caul is occasionally broken here, on Vancouver's East Side, the view through a rented flat, and when that happens the light opens a lane like a chute suddenly flooding with wheat-colored light that nothing can stop, that nothing seems able to stop, but then the side of a building does absorb that light, holds, and gives it back like a sheet of metal passing for stucco, granulated and desertlike light even though we are surrounded on all sides, including those we cannot see, by water. Note: New Mexico meets steam meets blare of ski lights and electrical wires meets orange lifts: autumn, then, gasp, mountains changed by sight of first snow.

☁

What mood is my cat in when she sleeps with her head facing down? Now she turns toward her paws for cover, as padded coverlets to her eyes, the better to turn herself into a ball of fur so as to turn away from the darkness of this particular day, only inadvertently to miss the flights of small birds through the window.

☁

You *put on* your coat in winter. You *pull on* your coat in autumn. Each act of self-cloaking determined by the season's mood.

☁

If we think of moods as zones, they can be global (those encircling regions that divide the earth from itself, from temperate to arctic); regional (war zones); or personal (comfort zones). Initially referencing the small range of temperatures that support human life, we now apply the phrase to latitudes of psychological risk or threat. Comfort zones apportion our survival, but the zone that we are in when we are *in the zone* seems to place us beyond extinction and into a separate sphere. A handful of almonds and a teaspoon of berries;

scentless forest flowers lining the base of dark pines; the immensity of the horizon line outlined in coral: when we're in the zone, everything illumines us, no matter the scale or the place where we are sitting. The zone we're in when we're in the zone is a place of eros and of pleasure, a G-spot or a sweet spot. It's the zone of unselfconsciousness and pure happening and yet I can't seem to figure it without an image of machines, of pistons set to proper speeds and puffs of steam emitting at perfectly timed intervals, of everything working and in sync: it's Fred Astaire in an engine room.

☁

When we're in the mood for creating a mood we call that "inspiration."

☁

Is it life (the happened) that pushes a piece of writing forward and requires that it be born, or mood (a tone, a hue)?

☁

In the case of certain moods, the minute you start to think about them, they cease to be moods. This could be a problem for this book. I mean, the point isn't to kill the mood.

☁

If sentences arrive on the doorstep of our consciousness while coming out of sleep, we should consider ourselves lucky for the chance such lines afford us to compare dream logic to mood logic.

☁

In lieu of grandiose claims and proclamations, I give you word portraits in the form of scudding clouds.

☁

Whether a particular mood palette is bequeathed at birth, or whether we can claim a new mood for new phases of our lives. Moods as the barometers of a personality's metamorphoses, or thermometers to their spiking, drop, or rise. "Hold still while I take your mood."

⌒

As a high school teacher, I used to try to make quizzes interesting by, instead of asking for a student's name and date, asking for their name and "mood."

⌒

All I had to do was to study those clouds and they'd tell me all I'd need to know about mood: three tufts match up with three gantries— we like it when things fall into place.

⌒

The temporality of mist's rising and sudden stilts made of snow. I should have painted it every day because I can't possibly recapture each shift of cloud cover's moods from memory. I just was never in the mood to write it down.

⌒

FREE TRIP TO SPACE TO CELEBRATE ANNIVERSARY OF SPACE NEEDLE. "I have no desire to leave earth's orbit, " I say. You reply, "Don't you mean earth's atmosphere?"

⌒

My friend, on the cusp of a first chemo session asks, is patience a mood? No, my reply would be, it is not; but why not? And if it is not a mood, what is it? It has always been touted as something more like a "virtue," and something that requires religious training to achieve. I think a good book should or could be devoted just to this idea of patience. And how im-patience came about (with its hint of imp-ishness) rather than un-patience or non-patience. Patience isn't a mood, it's an instrument, and impatience a flower that blooms with-out waiting for the season's change of heart.

‣

My teacher used the word "environs" as a verb as in "video art easily environs the viewer." Moods environ us.

‣

The way that music through headphones not only conjures a flood of memory but can't be shared even though one imagines sharing it—there's something more to it than the experience of surround sound (as mood), or the way a mode of thinking can overtake us until our brain feels as though it is cankerous in a particular place like a dark side of a moon (is that especially so when mood replaces desire?)—a "bad" feeling, as in "look how *x* helps them but never helped me!" Life sentences and clockwork narratives that keep us tethered to an old ax. Not the same as a *sad* feeling. Wordless. True.

‣

Cocoons as moods: those days on which one desires such—and weather—and drugs post-colonoscopy.

‣

Kenneth Goldsmith calls Facebook status lines "mood blasts."

‣

Fitful moods.

‣

What would it mean to have a mood repertoire? Extracting water-melon juice as a step toward making pink pudding: is this juice its essence? Or its mood?

‣

The mountains are only ever visible as shadows of themselves, or sometimes (as though they are) sideways bending, protruding trees—each case is the obverse of the other: either not-quite-there, or too-there; remote, or obtrusive. Each "view" of a mountain or the moun-

tains as backdrop, as rapidly moving accompaniment to our own
swift passage, as frontispiece or façade, as that which we, daily, face,
is the same in this regard: they look black but are really green; they
look blue but are really brown; they look smooth and unbroken but
are really ruggedly uneven; they appear preeminent but the play of
light, height, fog and cloud cover, drizzle and distance, only ever delivers to our gaze an outline, leaving us with the sense of a trick exal-
tation, an absence where we'd presumed the very shaper of our days,
a band and belt to measure our own heft against and rejoice in our
own lightness—the mountain range as something that can never
quite match or meet the idea of it, even as the reality exceeds the
idea. "A shadow of its former self" and, therefore, a mood.

☁

What's more difficult? Finding a language for our moods or for the
forms that clouds take? We want to say what a cloud is like—another
occasion for projecting our mythologies into the heavens, the heavens themselves, misnamed. Beyond reach (touch or comprehension?), but in view; near and far: just now, a sleigh driven by two
bears; paws, claws, curly haired monsters; typically, faces, poodles,
or elephants; or, no, that curling shape, a handle, a piece of pottery:
the sky seems the place where people most often stow their teapots
and fine china. Clouds *break* (a break in) without clank but rumble,
as do moods, but differently. How? Back-lit. Outlined. Dark. Not to
be confused with the plumes of steam from nearby factories. When
and why did we name them? The categories seem ridiculous. What
makes for drama? A cloud's departure or arrival? Its little shock. But
only if we are afforded a window. Impossible, perhaps, to make a literary form approximating clouds—easier to rustle up atmospheres
and call those moods. One explanation: water vapors in the air
around us in a gaseous state released from plants, trees, and evapora-
tion from bodies of water. Air rises and cools to reach the dew point
and the gas turns into droplets. There might be something to the
phrase "pissing rain," and now, it occurs to me, clouds as traces of
gassy expulsions: earth's farts turned into a field of wonder.

⌒

Moods aren't only containers of consciousness, wispy holders, but its engine: moods as motivators and directors of intention.

⌒

Moods bring to mind coloration, but do they adhere to our being's surface like dye to an egg? We speak of their washing over us or, like a storm system, moving in. More sepia in hue than Easter egg yellow or blue, a seeping sepia, our moods the effect of an accidental spillage in a photo developer's lab, when, out from the bath, grays and grades appear.

⌒

A mood never conforms to a human head like a halo or a helmet though we might wish for it to: to fancy ourselves being able to take one mood off, and put another mood on.

⌒

What do clouds do? They come. They go. Speaking of Michelangelo.

⌒

A friend of mine says people all his life described him as "moody," when he was really quite buoyant. "Moody" was a euphemism for what they didn't want to recognize as gay.

⌒

Moods seem to make us modern, modular (think sectional couch), and like washing machines with their settings and "modes." Moods might be commensurate with modes, not just their adjuncts: our moods are our styles and our ways. So let us turn to tables and gemstones and songs: all of which are set, like moods. Setting a tone is a felicitous image, and one, I trust, we are likely to enjoy, because a mood, once set, puts another mood in motion.

☁

There are sounds we need to filter out in order to hear what each other is saying and moods we can't afford to entertain—as if we had a choice in the matter. Moods forged at the threshold of our entry into language and to childhood.

☁

Mood awaits its own unanticipated form, a type of writing that holds a reader without holding her in place, that lifts and falls, and drifts and calls. It never leads, but carries a reader for this spell, then sets him down again.

☁

Moods are the clouds that constitute the banks of our being.

☁

Moods: something we retreat into or cannot easily get free of. A haven or a cage. A vastness with eddies.

☁

A night sky is not the same as a cloudless sky.

☁

Now I want a meeting place fashioned of differently angled and differently scaled inclined planes. Instead of sitting at long tables, each person lies on her back looking up. Each speaks without facing the other as at a campsite at nightfall, our documents in common: one star-stud or a spiel of constellations. Some are silent, while others carouse and carry on. Everyone murmurs on the verge of sleep.

# of clouds and moods

One of Sylvia's first acts when she rose was most significant. She
shook down her abundant hair, carefully arranged a part in thick curls
over cheeks and forehead, gathered the rest into its usual coil, said to
herself, as she surveyed her face half hidden in shining cloud—
—Louisa May Alcott, *Moods*

Clouds might be described as cottony, or like cotton balls, but such
obvious metaphors were nothing compared to plucking a real cotton ball from a paper bag, holding it first in my fist as if to weigh it
inside a palm suited to its size—no wider in circumference than a
quarter—then, applying it with small fingers, affixing it to a piece
of construction paper so as to pronounce the idea of sky. We always
evinced our makeshift skies this way in elementary school: in between counting and spelling, rambunctious recess, and reciting the
Lord's Prayer, a quiet cove would open like the setting of a table,
and we'd agree to concentrate on arts and crafts. By second grade,
we had advanced from blowing bubbles to gluing cotton balls into
place to form a scene—always with cloud-colored glue, the milk of
Elmer's moo—and the paper just as porous absorbed the cloud and
held it aloft. We had advanced from eating cotton candy or making
houses out of sugar to using the white and springy earthy plant stuff
to form a scene. Would you be a copyist? To craft what you knew was
wanted—cloud-form here, tree to one side, flowers beneath. Would
you be a realist? Or an imagist? I only knew this much: one kid used

but one cotton ball per cloud, while others massed them all on top of each other, pursuing three dimensions, and still another introduced so fine a tune as to tear the ball, and thus to teach us all that not all clouds are round or billowy.

Tearing cotton—who knew it could be done!—was startling for its sound: in the neighborhood of fingernails scratching a chalk board but pitched to lower decibels reserved for humming birds. It also felt strange: the same as I felt the time I bent my wrist bone without breaking it in a game of dodge ball. That same tensile pressure that a saw assumes when it is used to make a sound approximating music rather than to cut wood.

There were only two uses of cotton balls as far as I could tell—in the application of calamine lotion to poison ivy, year in, year out, the one pink glop the size of a button was cooling before it stung and later crusted over to form a pink film. Usually, some bits of cotton would stick to the surface of the viscous, minty salve, covering an itch, if only temporarily, with fake fur at an elbow or a forearm. I didn't just feel the cotton and its sticky dew in places—a daubing here, a dabbing there; whole afternoons would pass during which my whole being felt covered in a caul I wished I could break out of. Clouds appeared on those days to pass on the other side of a scrim opaque as a gold-fish bowl, with myself the goldfish, gill-less and unable to leap from the water. That was one use of cotton balls, whilst the other was in the creation of a sky.

I would gladly play with cotton balls rather than craft with them. Just to roll one around in my hand, or blow on it—it was so *other*, so like nothing; it was so *like*, like what? Like skin (a cheek, for instance), or bedtime, or maybe story.

Imparting a sense of cloud to paper through collarbone, shoulder, and down into fingers is perhaps all we can hope to do. The trick where cloud-writing is concerned may be to let a cloud pass through you rather than assume that observation alone is the route to understanding. Not to memorize skies along with lessons, but to let a memory rife with cottony sensuosity surface and depart. In her book-length poem, *The Weather*, Canadian poet Lisa Robertson broaches

clouds' age-old representational challenge by allowing different registers of language—meteorologic, newsy, metallurgic, adverbially
nominative, affective yet not in the least descriptive, diaristic and
diurnal, to crash, crush, float, interlineate and intermingle like so
many aqueous particles on a page: "Begin afresh in the realms of the
atmosphere, that encompasses the solid earth, the terraqueous globe
that soars and sings, elevated and flimsy. Bright and hot. Flesh and
hue. Our skies are inventions, durations, discoveries, quotas, forg-
eries, fine and grand. Fine and grand. Fresh and bright," she writes.

No wonder my earliest memory of clouds isn't of clouds at all but
of a medium for imaging them and a tactile lolling. Cloud tales rely
on a lexicon of elusiveness and capture, clouds as entities that lure
us with their indefinable definiteness: no sooner do they exert a pull,
like slow-moving, variously shaped tugboats asking us to board, than
they somersault into ungraspable dissolution. Students of romanticism, in particular, and its place in that foggiest-of-notions nation,
Great Britain, have much to say about clouds, producing fascinating
histories of weather routed most often by way of Luke Howard (he
who lent clouds the names they bear to this day in his 1804 *Essay on
the Modification of Clouds*), through Thomas Forster (he who helped
popularize Howard's taxonomy in his 1815 *Researches about Atmospheric Phaenomena*), through John Constable (the landscape painter
whose lifelong dedication to producing pictorially believable clouds
lent him the name "the man of clouds") to poets like Goethe (who
wrote poems to match Howard's categories) and William Wordsworth, the great poet of immersive skies in the Lakes, and John
Clare, who, wandering hill and dale, opted more often, despite his
anxious musings, for cloud ramble than cloud cover.

Clouds are "protean" and "ever-changing"; they are "ineffable
and prodigal forms," and "fugitive presences"; at once material and
immaterial, they mix something with nothing; they are heavy without being solid. For cultural theorist, Steven Connor, much of their
enigma "derives from the fact that they are a one from many phenomena of pure and irreducible multiplicity. A cloud is the temporary coalescence of a crowd of particles, each too small to be seen in

its singularity . . . [whose] most salient feature is to form themselves out of nothing and nowhere."

Categorizing clouds is one means of drawing them back into a perceptual grid that they are otherwise sure to exceed, and it's amazing that Howard's system of identification worked so well. His flat, low-lying "stratus," indicative of layers and sheets; the fluffy round "cumulus" approximating heaps or piles; the wispy, high-flying "cirrus," fibrous like hair, are simple enough for a schoolboy to master but capacious enough in their hybridized versions and admixtures to hold variability within their net, just as Howard intended: "The names for the clouds which I deduced from the Latin are but seven in number, and very easy to remember: they were intended as arbitrary terms for the structure of clouds, and the meaning of each was carefully fixed by a definition: the observer having once made himself master of this, was able to apply the term with correctness, after a little experience, to the subject under all its varieties of form, color or position." I think of poor Laennec, Howard's contemporary and inventor of the stethoscope, who tried, and failed, in similarly Linnaean fashion, to systematize, name, and then apply so as to diagnose the range of types of sounds that equally amorphous substances made inside the body's atmospheric chambers, bloody, breathy, wet or dry, fluent or blocked, bumptious or steady, cloudy or clear. Who's to say his science was any less exact than the index of indices that bred the weatherman? Howard's system, we might say, was capable of holding water.

There's no question that one experiences a kind of plenum the first time one discovers in what a user-friendly way Howard's codes match up with the peskily shifting nebulae that mark each day of our lives. But I think I'd fall into Caspar David Friedrich's camp if, like him, I'd been asked by Goethe to supply him with a set of illustrations for his 1817 essay on Howard's nomenclature. Richard Hamblyn, in his beautiful *The Invention of Clouds*, recounts how Friedrich resisted attempts to "force the free and airy clouds into a rigid order and classification"; for him, "the deep obscurity and impression of the clouds were valuable attributes in themselves." (Hamblyn, by the way, seems to fall in favor of Goethe who, he argues, enjoyed a free and open understanding of the ways in which science could inform, instruct, and be mutually inspired by art, whereas Friedrich was a kind of aesthetic purist).

The point, perhaps, is that a relationship inheres between clouds and human consciousness, human sense-making procedures, science, psychology, and aesthetics. Clouds, at base, must be conduits to *pensive moods*, however stormy, but there must surely be a difference worth exploring between thinking on or about clouds and thinking *with them*. John Constable again: rather than write that he was trying, let us say, to *depict* clouds, he described his practice in a letter, "clouds ask me to try to do something *like* them" (italics mine).

The relationship between clouds and moods is a quite complex one, both explicitly and implicitly. It's safe to say that clouds and psyches have gotten all mixed up: clouds literally affect what we take to be our mood states—and they do this in both conventional ways (rain clouds = sad mood) and singular ways (rain clouds = for you, in particular, good moods). But clouds are also available forms to project our feelings onto. Oddly, by mapping our mental states onto clouds, we seem to admit an affiliation between ourselves and clouds, but we don't go deep enough (or we only go as far as our egos will allow): we fail to admit that, riffing mildly on Shakespeare, "we are such stuff as *clouds* are made of and our little lives are rounded with a sleep."

Do outward cloudscapes map themselves onto our minds, interpretively? Do we seek in outward atmospheric marks the affirmation or negation of our own inner vagary? Or is our affinity with clouds an effect of our being made of the same stuff? Pollen, air, dust, water, and bits of volcano? Weather emanates from an unseen region at the center of the earth. No, it emanates from the bodily effluvia. No, it emanates from air in someplace far distant from where we're standing, breathing. No, weather is itself immanence. Moods and clouds pose a similar epistemological struggle: we are flummoxed by where either emanates from, yet we understand them as reciprocal. Who among us is not affected by the weather? Some will claim to be deeply, inordinately, and ever affected by it; others to be aware of how deeply affected they are only when the weather changes and a new surfeit of feeling suddenly becomes available to them. Much of what humans have built is designed to protect us from the weather—literally to "keep out the elements"—as though our major modern mode is to defend ourselves from it rather than invite and live with it. Hardly do we consider *our* ability to affect *it*, individually or collectively (consider, for example, global warming).

My cloud is my domain, my fantasy, my freedom, my snort of particles—"hey, you, get off of mine!" A mood "passes" or "lifts," as does a cloud. "I can see clearly now the rain is gone / I can see all obstacles in my way." Depressed, I'm under one; elated, I'm on cloud nine. Either I'm feeling the need to cry and am not able to, or it's the kind of day in which it wants to rain and can't. We are creatures of temperament, temperatures, and tempos. Our moods speed things up or slow things down, spatiotemporal, and temperospatial; like weather, clouds and minds are made of buoyancy or heft.

"As clouds race toward their own release from form, they are replenished by the mutable processes that created them. They drift, not into continuity, but into other, temporary states of being, all of which eventually decompose, to melt into the surrounding air." There's a hint of the anthropomorphic in this beautifully compelling composite of sentence unto sentence from Hamblyn that could describe the mutating evanescence that defines both clouds and moods. Depression, as I've suggested before, is in this sense not as much a mood as it is a refusal of mood's essential tendency to shift (not swing), but to seem to come from nothing, thence to disappear into the air. Eclipsing, overlapping, moods and clouds are layered transparencies, not just one mystic writing pad à la Freud, but many sheets of shapes divined, only readable from the side: viewing them sagitally, we glimpse a hint of color where light leaks in, orange, now lavender, now gold. Moods in this sense are like fore-edge paintings along the side of a book, some of which are visible only when the book is opened, others of which are visible only when the book is closed.

No matter how you look at it, clouds and feelings are mates, as intimately conjoined as soulfulness and music. "The ocean of air in which we live and move, with its continents and islands of cloud, can never to the conscious mind be an object of unfeeling contemplation," Luke Howard waxed, but there we go again with transposing land onto sky, consciousness onto its pre-formation and ground. So, too, Wordsworth disappoints me (but who am I to disappoint?) with his praise of clouds over clear blue skies: Ron Broglio references his *Guide to the District of the Lakes* where Wordsworth fancies how an Englishman should congratulate himself "on belonging to a country of mists and clouds and storms . . . and think of the blank sky of Egypt, and the cerulean vacancy of Italy, as an unanimated and sad spectacle." How well I remember the blue sky tease of a London sky or the hidden promise—the ever-tempting, swiftly vanishing patches of blue-maybe that daily mocked me on a visit to the Lakes. While blue skies might not be all they're cracked up to be— even I've been known to cringe at the compulsory (joyous) call of spring—Wordsworth should know better than to picture a blue sky as a vacancy or as a sky without cloud. No sky is uniformly blue—it only appears that way to those who fail to notice it. Competing hues of blue fracture and pulse without quite convulsing to make true blue behind which must be hiding yellow, red, or puce. Blue skies are clouds in existential dispersion, or cloud's aftereffects; inspiration to cloud's exhalation, they're slates for remembering coming clouds as harbingers of life's continuance and shade.

Perhaps taking their cue from Wordsworth, or perhaps based on the bad rap that clouds obtain (consider "brain fog," and "head in the clouds," "cloudy thinking," or "airhead"), a twenty-first-century online congregation called the Cloud Appreciation Society has formed. "WE BELIEVE that clouds are unjustly maligned and that life would be immeasurably poorer without them," their manifesto begins. The group, apparently based in Great Britain, enjoys a worldwide membership of cloud lovers entitled to badges, and if they wish, T-shirts, as well as a cloud-spotter app as guide to forty species of clouds for identification. While one might be tempted to compare the group to bird-watchers, the group's mission seems less about tracking clouds down or ticking off sightings while waiting quietly as a hunter with-

out a gun, and more about giving people an opportunity to post photographs of breathtaking cloud formations from around the world.
Clouds are "Nature's poetry"; contemplating them is therapeutic
and can save on psychotherapy bills, according to Cloud Appreciation Society's guide. Though no one goes so far as to claim to see the
face of Jesus in a Cinnabon-shaped cloud, a special segment of the
site is reserved for photos of clouds that look like things, reminiscent
of our tendency to take special pleasure in finding something akin to
life on earth in the sky if not exactly attributing such sightings the
power of a vision, a forecast, or an amulet. Pledging to fight what
they call "sun fascists" and "blue-sky thinking" wherever they find
it, the founders of the group craft an anthropomorphic mood pro-
jectile when they state: "We seek to remind people that clouds are
expressions of the atmosphere's moods, and can be read like those of
a person's countenance." Clouds are people too, I guess—such confusion pointing up our desire to tame the beyond-our-control mood
modulation of clouds, the degrees to which clouds hold us, and the
ways in which they perceptually beguile us.

Thus we move and shift, torporous or swift under cobbled skies,
alone with our thoughts, trapped behind bars made of rain, calm
as a low whistle inside snowfall's steady drift and quiet bound, too
bright to look beyond ourselves, we squint, hoping to be anointed
or better to be forgotten, lost or left alone in this shade, restless for
company. Cloud bubbles work this way, spheres of influence, ideas
in the air and so many of us in different locales breathing the same
air all the while groping in solitude and thinking we've struck gold
or gotten closer to a lunar surface. Today I find a fellow moon walker
in Mary Jacobus who makes the governing principle of my work on
mood seem plain as day. Early in her recent book, her newest study
in British romanticism, she articulates it: "It is this combination of
indeterminacy, space, and interiority that particularly interests me
here. Clouds, I want to argue, make us think not only about form
and vacancy, mobility and change, but also about the peculiar realm
of affectivity we call 'mood.'" Or, again, "Mood is like weather,
changing and unformed, yet always with us."

Mood work is suddenly timely, I'm not sure why, only that certain literary weather vanes have directed me for a very long time to follow them into spectacular day-vales, not of identification, but of difference. A line or sentence tolls, and I have to keep it nearby, not as comfort but as goad to my apprenticeship in writing light and time, discernment and gradation, graphite and water color, pure form and opalescence, stillness and movement, invitation and surround.

It's November 15, 1969, according to the entry in New York School poet of luminescence, James Schuyler's *Diary*. "The first day of winter, or if you want to get technical about it, the first wintry day," his entry on and for this day begins, and, with it, the recursive pause that casts the day as poetry. The sentence moves assuredly, parlaying *technos*, description and definition, at the same time that it requires a backward glance and self-consideration, since wouldn't "the first day of winter" be a more technical locution than "the first wintry day"? No, on second thought, the technicality in question is the date, and with that, whole worlds are animated in which we can question whether the start date and end point, the place where seasons begin and end are arbitrary or fixed, like the place where a poem begins and ends, the proper border between poetry and prose? What's the space between a day that *feels* like the first day of winter (or summer or fall) and the day that *is*?

The entry proceeds: "On Jane Street the leaves a skinny sycamore won't let drop shiver under a mollusk sky. . . . Close the dark blue curtains ('chance of a snow flurry or two'), hoard heat, hear a tired rhumba wind up the stairwell with the lank persistence of a kudzu fine, drink Schweppes Bitter Orange and wish you hadn't, feel kind of good, and write to Lewis Warsh, a Birthday Twin."

How often have I returned to that sentence, wondering how he got two verbs to jostle in such close proximity to one another— "won't let drop," and "shiver," and the way his syntactic mastery of pause leaves the taut resistance of the tree to resound like a hammer struck on the chord of "shiver." With what command of caesura (the sentence could be lineated as poetry) Schuyler's words work on me, making way to arrive at that word-cloud, a mere puff of a

consummate metaphor, "mollusk sky." Words have their own principles of condensation; where one word ends, another word begins
("sk," "sk"); ocean reflects sky just as sky becomes sea; sky and sea as
formed by the little phrase are linguistically and conceptually con-
sonant. In "mollusk sky," I'm reminded of Hamblyn's discussion of
the contributions of Thales of Miletus (ca. 624–545 B.C.), "a figure
widely regarded as the first real 'scientist' to have been produced by
Western civilization," who, "in maintaining that everything in nature was to a greater or lesser degree a modification of water, . . .
had voiced a fundamental truth about human existence: that we live
not, in reality, on the summit of a solid earth but at the bottom of an
ocean of air."

As this is Schuyler's diary, these are Schuyler's (pronounced "Sky-
ler's") skies, ("sk," "sk"), but no "I" bores itself as the be-all and end-
all, the starting point and end point of this most original of diaries.
Starting with "close the dark blue curtains," this diary's "I" emerges
as an effect (rather than preexisting progenitor) of an agglomeration
of languages: it's a mood portrait of sorts composed of weather reports and competing commands, self-reproaches, inward turnings,
outward correspondences, and the tired rhumbas of others. James
Schuyler's definition of a diary as I imagine it: a sketchbook for using words like iridescent watercolor; a daybook for writing light as
the most humbling, affective, and momentous of events.

In late June of '69 from the Maine island perch of the home of
painter Fairfield Porter, Schuyler observes "chokecherry that against
the light seemed in its bright darks to have light concealed within
it or resting in it." "Chilly saran-wrap and aluminum foil days with
beads of moisture condensing in them" give way to "a sunset among
Tiepolo clouds, blue and silver-white just brushed by gilt." In earlier spring of '68, landscape begets atmospheric forms and vice versa
with a "a mist made of clover," hawkweed that "goes in patterns like
a milky wave," and "depths of forsythia . . . brown as pancakes." In
January, outside, "a kind of gusty glare"; inside, a "trail of cold" that
rushes up between the poet's typewriter keys. In early summer, "a
radiance diluted and stabilized . . . [that] lay on the lichened trunks

of spruce and warmed away their woods chill." By mid-July, "Humid and cool (cold feet in wool socks) fog pressing in on the South Woods like a migraine headache, birds jabbering listlessly." Earlier in the year again, the "clear acid sulphur yellow" of that harbinger of spring, forsythia, requires invention—the words, "transpicuous light," which I imagine as a perfectly mood-toned cloud term: a cross between transparent and conspicuous.

Is cloud-writing tantamount to a writing that stays in the mind after the words evaporate (the way Schuyler felt upon reading *I Promessi Sposi*), and in that sense is cloud-writing a mood evocateur? Cloud-writing cum mood writing calls for an aesthetics of *chronos* and *chroma*, time and color, best conveyed by "lustre," a word coincident with both a period—etymologically, five years—and a sheen.

If I try to look at clouds, if I go looking for them, they can't affect me. Then they want verbs like perforate or punctuate, rather than arc and envelope and spread. Then they call for the stamp of aphorism: clouds: the Rorschach of the gods. They ask me to find correspondences: the steam rising from the body of a well-worked horse; fish flopping in the mist; your cigarette's smoke rings. Then the day is flecked particleboard as stiff as an attaché; the sky, spongy as discarded orange rind surprising the sidewalk, not yet discovered by ants. Today I look out for collateral images: a cloud is the outline of Queen Anne's lace having lost its intricate center, let it thrive in the middle as well as at the side of the road; clouds are notes hung from a sky-staff; the sky is the dull underside of an inverted pie plate stained with burnt cherry (only at sunset, never sunrise). If I go looking, I can only feel this mood state: *it was a day of exceptional promise.*

Skies most often take me off guard when I'm driving, rarely en route to work, usually when I'm coming home from the region called South County, Rhode Island, to the city of Providence. There are spring days on the back roads of a town called Slocum when everything at a juncture is dipped in platinum. My heart slows then and the car downshifts into a surreal gear. Then the thermometer fails to register as if the sun has become for that instant a blob of mercury and nothing is readable anymore. Something gets flipped for those moments granting a temporary peek at reality's underside: we are not where we seem; nor does it matter where we're headed.

On another day, I'm already on the highway—late February, early March—when what appears to be miles and miles of low-lying sheets of cloud announce a landslide in the skyscape—either that, or the Third Coming. We whip out our iPhones in a ridiculous attempt to catch it rather than contemplate it—I take ten pictures in five seconds. Luckily, I'm not in the driver's seat. Later, when we compare our shots to those of equally dumb witnesses on Facebook, I discover an accidental detail on the highway's other side: a white stretch limo speeds like an absurdist competitor beneath the looming avalanche of sky. We, too, sped in our lichen-colored Prius because I'm sure the effect of this cloud was to quicken our pace. Irrespective of any desire we might have had to be held by it, the highway seemed dead set on matching the breadth of its widening stripe, but, being out-lapped by the velocity of this form's vastly superior engine, had sloped irretrievably away from it until what seemed capable of overcoming us had vanished. I wondered what it would have meant to us if we'd been carless.

It's not nearly nightfall but after dinner when you close down one part of the day hopeful to open one more before sleep but uncertain of how or if you'll find the energy. You're moving on, only to be drawn more deeply in, slowing distractedly down, when something unlikely about the quality of the light inside calls you to go outside and look. How could something in the far off nearly night sky have penetrated so as to alter the shadows cast by lamplight of the objects in your rooms? You follow it, like the scent of cinnamon rising through fork holes of a just-baked pie, and find your neighbors out on the sidewalk too. It's so much better a group call than chasing fires or a siren, though, truth be told, there's a fire in the sky this dusk. Pink and blue and blue and pink then white: shuttered sky. The street's speed limit sign is perfectly illuminated, 25 mph, but few cars cruise the road beneath the tulip-arched streetlight easily mistaken for a moon. You expect to hear a crackling, but the electrical wires merely hang like lank cordons, or more deliberately, like cubist surveyors' lines. It's the trees that encroach wildly on this sky, their sinister and ruthless side hidden by day. And where are you in this? Nothing you did made this happen, and you don't want to be together with these other people beneath this sky, and you don't want to be alone. You've just been called by the day to suppose the alteration of a mood: to watch the nearly night sky slowly fold and unfold before tucking away a rose-colored dishtowel.

To have such mood-cloud experiences or vice versa is one thing; to make art from such encounters quite another. "We are the first generation to see the clouds from both sides," Saul Bellow's narrator remarks in his novel, *Henderson the Rain King*, "First people dreamed upward. Now they dream both upward and downward. This is bound to change something, somewhere." Henderson is a many-times-married millionaire who has this thought as he's purveying the clouds from the vantage point of an airplane on an ostensible pleasure trip: en route from the United States to Africa, he's accompanying friends who are traveling there on their honeymoon. A kind of lost soul who feels guilty about his wealth and insulated from the truth of things, he wanders in search of a purpose trying to get in touch with the world—for example, getting really earthy, he tries pig farming, but that fails. "I like the idea of clouds from both sides and some other things from both sides," singer/songwriter Joni Mitchell says after offering this précis of Bellow's character in her introduction to her 1967 performance of the song that she composed after reading this passage in the novel. Mitchell sounds like a school child or unassuming naïf when, in a quiet voice, she says, "I call my song, 'Both Sides Now,'" then proceeds to perform a master folk work of the greatest beauty and profundity.

Mitchell never finished reading the novel; she didn't need to. "Left up in the air," as she put it, she was inspired by the provocation of seeing clouds from both above and below by a character much like us all—in search of "what life was all about [to him]." The resulting song with its familiar refrain, "I've looked at clouds from both sides now," and its humbling conclusion, "I really don't know clouds at all," is as powerful in its harmonies and dissonances (in Mitchell's rendition at least) as it is roundabout in its ability to find its way into our hearts: each stanza opens not with a pronouncement of I-centered "feeling," but with a litany or collection of evocatively subjectless things, in the first instance, of shapes of clouds: "Bows and flows of angel hair / And ice cream castles in the air / And feather canyons everywhere / I've looked at clouds that way."

Some songs run like a vein of ore in our blood even if we don't realize how integral a part of our cloud makeup they are until we hear them for the first time again after a period of many years. Then, it's hard to know if the force of recollection makes the song feel suddenly essential or if it really is part of a mood-shaping bedrock of our being. "Both Sides Now" is one of those pieces of popular music inextricable from the year of its composition (1967), defining of an era, and belonging to listeners who were alive at that time—a song one calls one's own and that of others: part of a collective mood. The curious thing is that, affected as I feel by the song, it wasn't its author, Joni Mitchell's performance that I would have heard or grown up on. Judy Collins's cover of "Both Sides Now" familiarized it for American audiences, leaving me to worry that voice is arbitrary in how we come to incorporate a piece of music into ourselves: we don't know what we're missing, and we don't know what we have (which is not quite the same as not knowing what ya got till it's gone; more like, not knowing what ya got because ya didn't know what else there was to have).

By 1968, "Both Sides Now" enjoyed at least a dozen different covers in addition to Judy Collins's hit single. Listening to Collins's version is like revisiting a first crush, or better: it calls up the mood of macramé. Though I don't know how old I was or in what circumstance when I first heard the song (I was only eight years old in 1968), and even though Collins's version is polyester to Mitchell's leather, listening to Collins sing it again stirs me: I'm not sure if it's the force of will with which she sings the song that most moves me, or if there is something to feelings being buried in daisy-patterned wallpaper with matching kitchen cupboard contact paper because that's what the clavichord accompaniment reminds me of. The way her voice cracks on "at all" in the line "I don't know clouds at all" and later is held on these same words beyond what the song's tempo will allow: maybe this is what fills me with a sense of this cloud-themed song being about the ineffable something that is always held back in us, in me. Suitable to both a rainy-day pub or a daisy-filled meadow, pathetic as it sounds, the song features me stuck inside our family's powder blue Ford Falcon while harboring a yen for a red balloon–colored Mustang convertible.

From Bing Crosby, Robert Goulet, and Frank Sinatra, to, across the ocean, Marie Laforêt and across the universe, Leonard Nimoy, all manner of crooners chose Mitchell's song as their vehicle. The problem, though, may have been that they treated the song as a form of mood music rather than a song about the inscrutability of life, whether viewed from above or below. The song seems to want an alternative to illusion or disillusion as epistemological options; only a knowing being can admit that it does not know. In a YouTube video, Bing Crosby refuses to be a knowing being: he resembles a big-eared cartoon character; he sings the first few verses with his arms folded; he does something funny with his mouth—flicking its corners, as though groping for an absent cigarette; he tries, and fails, to achieve a Sinatra-esque understatement. Sinatra's rendition, meanwhile, might be characteristically swank, but he rushes through the song and jabs at the words like he's sparring with a silent partner (maybe Crosby?). By the time he riffs on the last line, interjecting the abbreviated "don't know it" between the eloquent, "I really don't know life at all," we're embarrassed by what feels like some form of oleaginous illiteracy. For Goulet, accompanied by marimbas, maracas, and tambourines, the song is all larynx. You can't sing a folk song as though you really wished you were singing an aria from *Il Trovatore*. This Adam's apple version may have worked to make the song popular as a form of bachelor pad Muzak or for housewives lost in the swirl of an alone-time afternoon drink with the shades drawn and the air conditioner set to high. Laforêt has French in her favor—all of those "ahjdge" sounds produced by a voice that is breathy, whispering, and deep make me feel she's at least trying to give an account of the song's mood, until, at the point of the second refrain, her voice suddenly morphs into a plaintive saxophone screech as though the song had put her to sleep and she were trying to ignite her own stupor. Nimoy's take is at first accompanied by an instrument that sounds like finger cymbals and later an electric piano busy with the sound of an unrelated computational analysis—a coy reminder of Spock? Straining to hit the song's high notes, Nimoy is forced at one point to speak the lines.

How many times will you listen to a song before deciding you've had enough of it? And if it makes you weep, does it only do that when you're in a particular mood? In Joni Mitchell's song, clouds are metonyms for that other great mystery of mysteries—not moods, but love, and by the end of the song, life itself. The lyrics say I used to see clouds in a dreamy way—as emblems of possibility; then I saw them as obstacles—they block the sun, they rain and snow; they're the reason I was stopped from doing things. Having seen them in these obverse ways, I still don't see them clearly or understand them any differently: I still fail to see them for what they are. The sentiment at the heart of the song isn't what makes me cry every time I hear it, and I think I could listen to the song endlessly, even in its bubble gum versions for the way they at least confirm the song's compulsive interest. There are three particular lines that always set my eyes to streaming: "Moons and Junes and Ferris wheels / The dizzy dancing way you feel," and "To say 'I love you' right out loud."

I'm with a friend and we're trying to tease out the difference between Joni's and Judy's songs. It's a canyon-feathered cloud of a day of recurring "deliciousness," which is one of my friend's favorite words. It's summer time, and we're breaking from a vacation day of bike ride followed by swim to make huevos rancheros with fresh eggs and lime, sweet onions and goat cheese feta, dashes of hot sauce and midday white wine. It's a most-splendid-of-days sort of day of reading quietly together and embracing—my friend is a perpetual hugger one of whose favorite utterances is "yay!" She's the sort who smiles with her whole body, and whose insistent delight fills the room. In spite of this, I'm aware in me of a muted darkness, a vacancy that wants to turn nimbus, cut off, suppressed. For the whole course of our conversation, in my line of vision: a pink flamingo in the nearby woods—a souvenir of another Maine-transplanted, originally Floridian, friend who disbanded her collection of Maine-staked pink flamingoes one to each of us before she died last winter. The friend who is my afternoon companion is her partner.

We compare the versions: Joni's guitar isn't simple "accompaniment"—this is a major difference. She lets the guitar sing in parts and waits for it to finish before following up with her voice. Joni's range, even inside a single word like "lost," is as warm as it is piercing, dense as fabric, transparent as glass. The song, in her handling of it, is made of carefully graded steeps and moment-to-moment shifts of octave, of clouds as seen from above or below: her singing, a lonesome calling across a cavernous sky like a yodel. Still, it's Judy's, the popularized version, that moves us both—I'm crying; my friend is weeping.

The dizzy dancing way you feel at a carnival: it must have a nostalgic pull for me, evoking a way I think I once felt—it's so recognizable—but a mood one can never know again (it's attached to youth, and 1968). As for saying I love you "right out loud," I must be moved by the blunt blatancy of it, the way speaking truth and welcoming consequences feels. Then there's a global feeling that could account for both our tears: the way that a mood-evocative song like this confirms existence. Decades ago, you were where it was when it entered your airwaves; it reminds you of your having been. Now you are here, but someone else is gone who had been with you then, just as you will someday disappear. Can a song retain within its reverberating strings and chords a trace of all its listeners? "Both Sides Now" moves us as a remnant of our own existence—that existence needs confirmation is enough to make a person cry, but it met a local grief for us that day as well. We were saved by its cloud occasion in this: to share what we could neither adequately say nor feel.

Each day is differently clouded, incrementally mooded, with no two days, clouds, or moods alike. It must take a perceptual sleight of hand to insist as we all do that days and moods repeat themselves as one vast unvariegated cloud form, end to end. Constable lent us the verb "skying" for his practice of regularly, systematically studying the skies and the forms that floated therein, clouds, in his words, "non-captive balloons." Presumably, he was provoked to undertake a more thoroughly scientific observation of the skies by critics of his art who found his clouds to be unbelievable, too weighty, prominent, or labored in their execution. Study doesn't have to yield exactitude, however, and the aim of Constable's cloud art needn't be confused with accuracy. What's most memorable to me about some of the cloud studies of Constable that I've seen is their size: the incalculable vastness of a cloudscape carried over into a six- by ten-inch canvas. A cloud is something you can carry around in your compact, a small mirror reflecting one part of your face. What Constable arrives at in his meteorological study of the clouds isn't an impossible truth but an intimacy. It's "a specificity that exceeds classification." It's a personal largess of micro-scatterings, the "scraps and bits of paper" on which he made his observations on clouds and skies never formed into a lecture as he had intended before he died, but transmuted into brushes of strokes of color and lines.

The painter Henry Fuseli and others have found a persistent mood in Constable's work—"the feeling of the threat of oncoming rain." (Fuseli said the landscapes of Constable "made him want to call for his overcoat and his umbrella.") Among Constable's "favorite weather stories," though, according to Ron Broglio, was the "glimmering landscape just after a rainfall." One of the things that makes clouds and moods so hard to illustrate, study, or pursue is that they're ever-changing, which might be why we like to project physiognomies into our skies, to imagine the sky itself as guard or guide, more rigidly, a face, and steadfast: to conclude that if the sky is above us, it must be looking down on us, and if it's looking down on us, it's watching over us, helping us to know what to expect.

The last time I saw the man in the moon, I was living in Buffalo— that place where the only thing that's predictable about the weather is that it will change dramatically several times in the course of a day. It was there that I learned about snow "squalls," lake "effects," and

wind tunnels; of the difference between snowy sidewalks and drift-
ing tundra; of icicles and ice cycles; of the imprint of feet into absolute quiet and the disappearance of a limb into a bank; of sheets of
translucency and gray bright days; of beading glare and sun-washed
showers; of the bliss of interiority and the necessity of inward dwelling. I was in my midtwenties, in graduate school, when I remember suddenly recognizing the genderless and kindly face, neither
bald nor in need of hair—more serenely above it all even than the
Buddha's—what everyone had referred to as "the man in the moon."
As a child, I think I had pretended to see it, and it's still not clear to
me if its visage is only available in moments of absolute innocence or
in moments when the table set by experience is wiped clean by what
one is coming to know, how one is coming to learn. You might say
atmospheric conditions have to be just right for the moon to show
its face as such, but I'm not so sure about this. I only saw it then, one
night in Buffalo, in the same era in which I grew into a habit that
I have never revisited since: of spending at least two hours every
morning—longer on weekends—lying on my back and staring at
the blank ceiling "thinking" before getting out of bed.

Those were precarious days for me of ambient anxiety; of feeling determined; scared; in love—with both ideas and persons. I
wouldn't say my bed-bound mornings were commensurate mourn-
ings: I don't think I was depressed. It was rather as though I were
testing the pliancy of a hammock of solitude. I do believe studying
the persistence of blank ceilings helped, for me, the skies to open
up. Seeing the man in the moon filled in the patchiness of my own
face; it filled me in with a feeling of "now I get it!" alongside a cock-
eyed whirligig of a sense: now the moon was eyeholes punched into
the silk siding of a circus tent awaiting the assumption of your gaze.

Of course conditions inside and outside need to be right to make
us want to look upward rather than forever forward into our day's
routines. What is that day on which the space between earth and
sky is eclipsed, abridged? What gives us the feeling that skies move
while earth stands still? Sheets of rain today are replaced by sheets
of sun: the storm is not yet over. During the same period of my life

of starting days by staring ceiling-ward, I discovered the paintings of
Western New York artist—the electrified, vibrating patterns—the
sky and landscapes of Charles E. Burchfield (1893–1967). While my
discovery of Burchfield in the 1980s felt accidental, today I reen-
counter him by way of a wall calendar encouraging the contemplation of one painting for each of the twelve months of the year assembled by the fairly new Burchfield Penney Art Center in Buffalo.

"His personal symbols and abstract conventions formed by sin-
uous lines and geometric patterns evoke sound, season, movement,
and mood," the calendar's compositor explains.

The watercolor painting I have described above is titled *Clearing
Sky* (fig. 2) and was painted in 1920; it is used to illustrate February
in its calendar context but to me looks much more like March. For
March, they give us one of Burchfield's wallpaper designs, titled *Red
Birds and Beech Trees*, 1924 (fig. 3). While the ground is interrupted
by a startling pink from time to time, harbinger of spring (?), the
mood of the work to my mind is Poe-esque fall. Or maybe March in

(2. *But look: the sky has changed again without me in it*)

(3. *to let a cloud pass through you*)

Buffalo. The sky is lavender-brown; its clouds are eggshell thickly
outlined in blue. Red bird-shaped blots fill a forest of trees made of
clouds: the trees these birds inhabit—are they cardinals or wood-
peckers whose red beaks have assumed the shapes of their whole
bodies?—don't just reflect the clouds in their slick trunks; cloud
draped, they wear them and deeply shade first flowers colored black.

(4. *"transpicuous light"*)

A small pool of red could be a poisonous tarn, or a rabbit hole to season's future's past.

A painting called *Early Spring* (fig. 4) reads more like a cathe-
dral of snow foregrounded by a riot of sunflowers or daisies: it
hangs above our calendar for April. This is a late career painting for
Burchfield—it was painted in 1966. Here, too, there are sky-clouds
and ground-clouds: those in the sky are underlined in yellow; those
on the ground are overlined in black. If this is snow, it is not melting but ascending, just as pines spire upward, suffusive in their aiming high. In *Gothic Window Trees* from 1918, Burchfield shows trees
as frames for clouds, as mediums for diffusing and channeling, not
merely, blocking light.

My favorite of the bunch is *Afterglow*, 1916 (fig. 5), its date of
composition making me wonder how the war figured for Burchfield,

(5. *you've just been called by the day to suppose the alteration of a mood*)

making me reconsider the painting's "glow" as the aftermath of a violent burst. Parts of the sky have been cut out; other parts, erased. Vast orange clouds are behemoth, winged beaches to islands in a distant sea. Purplish-gray tubular mauve-ite arms and legs of sky reach in, alarmingly, while a crescent moon-sun, outlined in brown, resides, we don't know where. Beneath this mostly sky scene, a house settles with a keen of vacancy. It's not just that something is happening out-of-doors that we're not present to, a seismic mutability we once again have missed. It's also that there are no people in these scenes, only human consciousness as vacant dwelling.

Burchfield may be best known as a painter who gave form to earth's vibrations. His vibrating patterns are his signature trait, and they can make you feel your own quivering when you encounter them. Where clouds are concerned, he dares to outline them without holding them in place. He doesn't lasso clouds but expands his canvas into them. He demonstrates how, by treating an outline as a broad band, a spectrum really, an amorphous entity can become visible to us—enlightened—rather than contained. By intuiting rather than applying cloud outlines, he illuminates inversions from both sides now of clouds never static but spreading like pigment to paper that spills only so far as its saturated edge.

So my pursuit of mood cannot have as its aim to capture, to make mood wriggle and writhe; if we're to write clouds, we must hope to liberate them in the same measure that they liberate us. In search of mood hints, I must let clouds act upon me: open and oblique. I must start with tools like cotton balls and construction paper, gray then black now red.

gong bath

All I had said was "mood" and "sound" and "envelopes," in response to the question, "What are you working on?" when a friend of mine invited me to a group event, or an individual experience, I wasn't sure which. It would require twenty dollars, she said; it would last for about an hour. She said she thought I'd really get a lot out of a "gong bath."

Immediately I pictured a take-me-to-the-river experience. I think I thought a midwife might be present. I needed to know if nakedness was a requirement, or if a bathing suit was optional. I imagined a toga, or endlessly unwinding winding sheet. The water would be turquoise-tinted and warm—bathtub-warm but bubble-less. Everything would depend on my willingness to go under—to experience a form of suspended animation. No doubt sounds would be relayed to me—underwater healing sounds—to which I'd be asked to respond with my eyes closed, all the while confident I would not drown.

Then I remembered how my mother was ever unable to float and how her fear of water fueled her determination that my brothers and I learn to swim early on. My mother can't swim, but throughout my childhood she wrote poetry in response to the call of a nearby creek that she studies and meditates near. I maintain an aversion to putting my head under water even after I do learn how to swim. How can I ever push off or dive deep if my mother cannot float?

Swimming won't ever yield the same pleasure for me as being small enough to take a bath in the same place where the breakfast dishes are washed. No memory will be as flush with pattering—*this is life!*—as the sensation that is the sound of the garden hose, first nozzle-tested as a fine spray into air, then plunged into one foot of water to refill a plastic backyard pool. The muffled gurgle sounds below, but I hear it from above. My blue bathing suit turns a deeper blue when water hits it, and I'm absorbed by the shape, now elongated, now fat, of my own foot underwater. The nape of my neck is dry; my eyelids are dotted with droplets, and the basal sound of water moving inside of water draws me like the signal of a gong: "get in, get out, get in." The water is cool above and warm below, or warm above and cool below: if I bend to touch its stripes, one of my straps releases and goes lank. Voices are reflections that do not pierce me here; they mottle. I am a fish in the day's aquarium.

The gong bath turns out to be a middle-class group affair at a local yoga studio, not a private baptism in a subterranean tub. The group of bourgeoisie of which I am a member pretends for a day to be hermits in a desert. It's summertime and we arrive with small parcels: loosely dressed, jewelry-free, to each person her mat and a pillow to prop our knees. We're to lie flat on our backs, we're told, and to try not to fidget. We're to shut our eyes and merely listen while two soft-spoken men create sounds from an array of differently sized Tibetan gongs that hang from wooden poles, positioned in a row in front of us. Some of the gongs appear to have copper-colored irises at their center. In their muted state, they hang like unprepossessing harbingers of calm.

~~~~~~~

At its furthest reaches, science's mood is poetry, at that point where it gives up on controlling the things it studies, agreeing instead to a more profound devotion to spare sounds whose tones the mysteries of existence brush up against asymptotically: the rustle of pages weighted with results, the fluttering of questions pondered in obscurity, the settling of a log on a forgotten fire, the hiss inside the grate. Even in its earliest incarnations, the science of acoustics turned to

water as its scribe by dropping a pebble on a liquid surface—plunk—
and watching the rings around it form. So, too, mood finds a home in
circles and widening gyres: the geometry that accompanies mood—
whether fore-, back-, sur-, or gr- is "round." And now these gongs
waiting to be struck are also ringed, from darkest center to shim-
mering edge.

I have noted before that I am easily seduced—ever in the mood
for love—though I'm not sure that makes me a quick study. I'm a
ready convert to any religion, keen to smuggle its riches into the
waters of a deeper understanding: this art. Which might explain
why my first gong bath was so affecting, but the power of suggestion
is only part of it. If sound's amplitude is full enough and the roof
beams not too low, if the human subject is surrounded on all sides
by sound, she really has no choice but to give in to it.

Our guide explained that the sound of the gongs had the power
to fill up every particle in the room until a bath of sound was
formed—a gong bath. It was true: the sound was so highly resonant
and painstakingly slow to fade that I began to feel awash in it. For a
person who hates to swim, I was amazed by how the more the sounds
filled the particles that made me *me*, the more I felt that I was living
in some blissfully underwater place without the need to come up for
air. Sometimes the sound was bowl-like; other times, it was bell-like.
Think of the sound achieved by running your finger around the cir-
cular edge of a glass, but the glass is made of felted metal or of wood.
Sometimes the sound was snared, faint as the needles against paper
on a lie detector test, or birds' feet sticklike in snow. What sounded
like water pulled forcibly over pebbles made me feel my body was
literally raked. Other times the sound was a booming trundle, loud
enough to liken you to early theatergoers who fled their seats convinced by the screen's illusion of an oncoming train in 3-D. But you
stay your course, not knowing what's next, only that the gong's most
powerful effect has been to enliven one part of you while making
another part supremely groggy.

I know it will sound like I was tripping if I say I felt as though I
was dropped down a watery chute inside a gong bath. The sounds

slowed things down to the point of a drugging of my inner voice: suddenly that voice was the cab of a hot air balloon that I would need to
climb up into to enter should I ever feel the need to return to it. Is it
possible for the mind to revert to pure sound? I began to have a feeling I'd never known before: my eyes weren't rolling backwards into
my head—this wasn't exactly an ecstatic state; behind their closed
lids, my eyes felt as though they were sliding to either side of my
head. This must be what happens to us when we die, though I wasn't
for that moment afraid of dying.

<center>~~~~~~~~</center>

The lover's discourse—any word uttered by the beloved—takes up
residence in the lover's body and rings there unstoppably. This pang
that requires Roland Barthes to halt all occupation he calls "reverberation." Without the aid of microphones or speakers, the sound
of gongs materializes and reverberates in the supine body—for my
own part, I felt sound enter through the palms of my hands and the
heels of my feet. In the concert hall, a cough or sneeze, whisper or
crunch is a too-ready reminder of the body of our fellows in the
room. At a rock concert, we maybe sway or sweat together in a half-
high haze but careful still to keep the edges of other bodies ablur;
we pitch our tent on the edges of group oblivion. In the gong bath,
other bodies are nodal points that sound bounces off of. I felt sound
bounce off the body of the person next to me, onto me, and on down
the line; I felt it in my stomach like a pang.

Here we might want to pause to distinguish between auditory
hallucinations and auditory hallucinogens, with the gong bath a
form of the latter. Was I letting myself get all New Age kooky, producing a form of socially acceptable psychedelia that has no basis in fact? That sound can affect the central nervous system goes
without saying. That sound can therefore be harnessed therapeuti-
cally to allay pain or alter the course of a disease has never been
the drawing card of modern Western medicine. A little research can
go a long way, and a student of mine once made me aware of pre-

scribable sounds, or "audioceuticals." Vibroacoustic therapy is often discounted as simply silly, along the order of overly priced vibrating easy chairs, until someone gives a sound massage to a person with Parkinson's and finds that circulation is enhanced, and rigidity decreased. White noise as a treatment for ADHD, vibrating insoles to help the elderly maintain balance, or the space age sounding SonoPrep—a skin-permeation device through which a blast of low-frequency ultrasonic waves opens a pore in the skin in lieu of a needle—suggest territories we've barely begun to broach. Though neither I nor anyone I know has been offered a noninvasive therapy tool that can liquefy tumors of the prostate and the breast, or sonically bore a tiny hole into an infant's deformed heart valve, the sound technology and its practitioners apparently do exist.

What's this got to do with mood? Applying sound to mood is not my method; I want to make mood and sound sidekicks, to consider them in sync, then see what emerges from that thought experiment. Oceanographers tell us that sound moves faster in water than it does in air, but isn't air part liquid? They say they can measure qualities of sound that are impossible to hear. They observe that sound pushes particles together and pulls them apart and that sound is the effect of a material's compression and expansion. When they add that the speed of sound in water is dependent on night or day, temperature, weather, and locale, I begin to feel I'm in the realm of sound with mood. So, too, when they describe a dolphin's "kerplunk" as a slap of a tail on water to keep an aggressor at bay; when they note a whale's "moans, groans, tones, and pulses," and a seal's underwater "clicks, trills, warbles, whistles, and bells," I begin to glimpse a mood, part-sea.

A philosopher steps in and says the body itself is a skin stretched over resonant matter beneath. We are our own water-filled drums of emotionality and indigestion, of sounds and moods. A poet parts ways to say that water is sound; sound creates moods; all mood is aqueous sound.

It's the feeling a gong bath gives of encountering sound beneath a threshold, submerged, and then absorbed that makes me ally sound with mood as liquid. The gong bath doesn't affect my mood—it's the model for a mood; it is a mood, and it can't be reproduced. It says that mood and sound meet at the place of touching. Sounds touch me, and mood is the window of allowance, wide or narrow, to let sound in: my moods are equivalent to what I let myself touch, and be touched by in turn, but also what I have no choice in the matter of being encased in. A tongue stuck to a cold pole; bare feet in mud. The bare of your back; the sting of my words. If I were a cat, touch would create a purring machine; if you over-touch me, I swat. Give us this day our daily sounds. How conscious are we of our ability to create our own soundscape exclusive of earbuds? How will you tune your day? What will you tune into with no instruments at your disposal but your whistle and gait?

~~~~~~~~~~

The gong bath gives the impression of being touched at all points from all sides, the body contiguous with air as substance. It's a bit like the effect a mime gives off when he produces the illusion of an object world that he touches and that we also feel, except in the mime's case, the perception depends upon the world as soundless.

Lest I seem to idealize my twenty-dollar experience, I should note that fifty minutes was way too long a time for listening to gongs. Five minutes would have had the same effect, but the gong players wanted to give us our money's worth. Every gong bath since my first one has left me cold. They've really flopped. The second one I attended was on the far other end of the continent from the first, but the guide, it turned out, had trained the people back in Providence. The room was too small, and everyone felt nervous. Nor was the atmosphere improved by the suggestion that we could have the same experience if we only bought the leader's homemade CDs, which he stacked and unstacked in a sad little pile at the beginning and the end of the class. The third and final bath was headed by an overly self-conscious woman who talked more than gonged, who sang songs whose lyrics likened humans to totemic animals, and who called upon the healing winds. It was cold in the room, and some people wrapped themselves in so many blankets that they appeared as a row of impenetrable pods or middle-aged campers devoid of starlight.

What happened to me in this gong bath is that I never got past the all-too-probable tendency to supply an image to every sound I heard, even entire narratives. Though the images were as unconsciously imbued, inexplicable, and private as those one experiences in dreams, I remained a foreigner stranded upon a shore and not a bather, immersing down and in. The images were dark: a boy shivering in his coat before drowning; my open mouth attempting but unable to pronounce Jean's name. There was a toucan and a typewriter, an avalanche of marbles, a body encased in wax. Having stirred up some unpleasantly tinged flotsam and jetsam, the gong bath left me feeling bereft, unlike great music "that move[s] us . . . because it is expressive of sadness," not by "making us sad." Sad music puts us in an exalted mood, rendering us capable of experiencing the expression of sadness.

In order for a gong bath to work, sound has to obliterate language for a spell so we can touch mood's casement, its resonant shell. We have to be coaxed by sound to suspend our image-making tendencies, even if pure mind like pure sound is impossible. But why should we try? After my first gong bath, I was convinced the phenomenon was going to become the audioceutical fad for twenty-first-century Americans. It could join the ranks of our half-understood borrowings from traditions not our own, providing an opiate to the all-too-comfortable classes, a soother to a whine. My prediction was a way of denying that I was in search of something, of an experience, deeply felt, and not just an observer doing fieldwork. I wanted to be invited to go under while you provide the sounds, to shed anticipation and bathe in curiosity, alive for a spell in the day's aquarium.

sonophoto: boy, screaming

It must have been the absence of a flash at dusk combined with the relative shade of my backyard garden that yielded an entire roll of mostly blank birthday pictures that year. The conditions, combined with a cranky camera, weren't going to make it possible for any images to materialize, which must be why I received the one photo that partially did turn out—of four-year-old Kolya screaming—as a tonic. He's a trace in this soliloquizing picture, and yet so vividly visible: there's a hue of a bare leg (it was summertime), and a broad white band of a striped shirt against the gray-green dark. Most especially, there's an audibly visible face, the angle of whose eyes shows the boy to be smiling—mischievously? triumphantly?—at the same time that he screams. His whole ghostlike figure tilts as if spinning inside the vertigo created by his cry, or maybe it just made the photographer woozy while the boy himself remained protected. This is my favorite detail: he's plugged his own ears with his fingers (fig. 6).

When Kolya was not that much younger, his father, my friend, Arthur, chalked up the extremity of Kolya's antics to the "terrible twos," but I never went in for that. The energy of Kolya's fervor was unique, or at least strictly his: he was his own heat-, light-, and energy-generating machine, which is how this photograph came to be. Against the absence of light and the possibility of a reproducible image, Kolya produces what physicists call sonoluminescence. Investigators still aren't entirely in agreement about how it happens, but they have observed emissions of regularly pulsing (in picoseconds)

(6. the tenor of these times)

eerie blue light occasioned by high-frequency sound waves (just beyond the range of human hearing) as they bombard an air bubble
suspended in water. Imagine finding ways to measure what goes on
inside of tiny air bubbles as they react to sound, heat up, collapse,
and emit a different form of energy: light. To say that Kolya's scream
was sonoluminescent is far from scientifically accurate, but what he
occasioned with his voice is more than metaphor, the literal projection of his being as an emanating source of temperature and tempo,
of atmosphere and mood.

Kolya was no more trying to *convey* something with his piercing
shriek than he was communicating anything in his two-year-old feats
and tests, as when he'd impulsively drop objects from our rooms out
the third-floor window. Could the snow-cone napkins he'd attached
to each item as imaginary parachute work without his also supplying
strings? Of course not, but the velocity of the plunge, the sound of
each object on hitting the ground, its relative intactness or scatter,
its liability to bounce, and the impossibility of return to Kolya's body

as projecting surface met him in the belly of a mood. It gave the form of an afternoon hobby to libidinal pleasure.

Those aren't the impressions of a cat's paws left on the hood of our car. They're the size and shape of a little boy's sneaker. He's running headlong into the wind of his own breath—nothing hurts here, he hurtles before he leaps to grab onto a cable wire he wants to climb.

Now I wonder about the nature and quality of his photographed scream. How high was it pitched and to what extent did it muffle or mute the other ambient noises of that early evening background scene?

I once purposely left my camera home on a trip to an unfamiliar country, deciding to record nothing but a range of sounds as remembrance of the places we'd visited, from those that created an accidental ambiance—a voice suddenly singing in an alleyway—to those that alerted me to the distinctness of the place—the uncommon, to me, clapper that signaled an approaching train, cowbells on a hillside, voices in a restaurant, and especially the tune of clanking silverware floating through apartment windows at midday. I was convinced somehow that the sounds would offer a truer sense of the place than any two-dimensional picture plane could. If I reassembled the sounds when I got home, could I create something out of view about the place but immanent?

A video couldn't have accomplished the re-creation of an atmosphere since sounds would be subordinated once again. Sounds by themselves, sounds and nothing but, have the effect of replacing us elsewhere. They bring us back, far, far back, and they bring us out and over to land in an extraneous zone, for sounds untethered are both reaching and vault-like, they emit a volume and an architecture, a substance and a shadow: but here I fall back on the language of images once again.

If we must have images—if we are never without images—and I do love my photograph of "boy, screaming"—then how about a *sonophoto* or an *audiopic*? Neither a moving image nor still images with wallpaper musical playlists made to match the images after the fact, a sonophoto asks to be a still image with, let's say, an accompanying audio chip as a record of the sounds that were synchronous with the subject caught that day at the very moment of its catching. For every predictable occasional shot, there'd have to be an unanticipated suite of sounds and voices, burps and bumps to mark the pic as such and bubble up from it, and merge with it, to surround it like a sphere. Audiopics would register the tones that collectively jostle a moment's mood into being. Photographs might seem less bereft then. They'd not be lent a missing sound—don't get me wrong: the idea wouldn't be to animate the images we seek to reproduce but to let them emerge and recede into their native sea of sound, albeit necessarily clipped and cut.

Here's where a challenge would have to be met because what would determine how long or short the photo's accompanying sound recording should be? A split second wouldn't do for a palpable mood song. You'd need at least several seconds for the sounds to be meaningful—or maybe not. Maybe the whole point of sonophotos would be to train our attention differently: to give us what we otherwise can't hear as distinguishing and vital and exceedingly short. It's the nature of sound to vanish, but memory seems to require the sense of a sustaining chain.

Would such sounds fizz around a photograph like tiny bubbles or create one large gelatinous frame?

~~~~~~~~~~

Unintentionally, at first, infant mouths create bubbles from tiny pearly types that dribble indistinguishable from food or milk but resonate with sound without the privilege of vibrating teeth. Lipbuzzing bubbles unleash delight, a redounding tickle on the surface of our tongues. In not too long a time, a baby practices making bubbles form-fitting to the mouth's O between the lips. A translucent lid as interface, it intermingles with a yawn or a husky baby cough, expellant, until it bursts then starts again. A baby's eyes at this time, afloat in their sockets, more liquid than sighted, are not so different from a bubble in the mouth. To introduce soap into this scenario seems like a civilizing leap—bubbles as objects of detached contemplation, or as playthings that we can take or leave: can you balance one on a stick? Can you pop it before it drifts off, can you bear or unloose its relationship to you?

"Froth," according to an Aristotelian model of temperament and physique, "is the euphoric counterpoint to black bile," but the four humors only allow for choleric, phlegmatic, melancholic, and sanguine. This photograph of my boy screaming, though, could serve as a woodcut illustration to a type of "mood euphoric" in which the figure has surrounded himself with a bubble of light. How long did he sustain it? That scream. Was it sharp and fleeting like a yelp or long and lasting far beyond the boundaries of a measured exhale? By the looks of Kolya's self-constructed earplugs, it was exceedingly loud and high-pitched. Was the whole point of the scream a test of how far he could throw his voice for how long?

Much as I might wish to resist the associations, I know it's difficult not to hear a call to gender inside these sentences. As though length and endurance always have something phallic built in. The great theorizer of voice in cinema, Michel Chion, points to how a gender emphasis inheres in the two terms in English for wordless cries—"we tend to call the woman's cry a scream, and the man's cry a shout." In infants, I'd argue, the terms are unilaterally applicable, one set of baby lungs and needs being as fiercely undifferentiated as another. But the adult male shout (in cinema and perhaps imitatively outside it) marks and delimits a territory of willfulness and power of however variable duration (see Tarzan's liquidly pulsing yodel); it is "centrifugal and structuring," whereas the female scream is a cry in the face of death: limitless, "centripetal and fascinating."

Such cultural performances of screams—and with them, emotion and moods—must explain why we find a deep voice in a woman sexy but not a high voice in a man.

Or why some people (myself included) are moved to tears by the dulcet delicacies of an all-boy choir but not by the collectively beautiful voices of a girls' choir. A girl's voice isn't expected to mature, and each generation of women has a different set of techniques for keeping their voices childlike. Innocence is here today, gone tomorrow in females; it's as much a state as it is a thing that can be relinquished or taken, and it is governed by our relationships with men. Whereas male innocence bears the built-in pathos of an inevitable loss that has to do with a man's relationship with himself, marked by the deepening that will make the boy's voice into a man's. Male bodies and male voices are more freighted with emotional investment, then. There's more understood to be at stake in their shifts and changes.

How about the sexiness or lack thereof that attends a voice's gender and pitch? There's only a seeming opposition or asymmetry at work here. We hear in the high-pitched male voice effeminacy—read, emasculation: it's a no-brainer. But a deep voice emanating from a female body actually isn't read as masculine. That doesn't explain its charge. A deep female voice—husky, sultry—signals a drinking, smoking voice, it's an easy, well-lubricated sucker of a voice that bespeaks "availability."

It's hard, maybe even impossible, to create a taxonomy of types of human cries, then map them onto the feelings that generate them, or the moods they create. Was the Beatles' "Twist and Shout" meant to celebrate a genderless orgasm in an era of mass protest? Screams can evoke both pleasure and pain; we emit cries of joy and cries of sorrow, but we apply the verb "to cry" more commonly to sadness. Gasps, gulps, and whimpers. Hysterical squeals.

Nothing equals the street opera of pushcart peddlers, the market voices of Sicily or Seattle, say, though in Sicily they meet my ear more plaintively, they seem more deeply about grief and less about sales. Sicilian vendors do not bark or scream; neither do they yawp or shriek. The mostly male voices *cry* into a vibration held long enough to draw me to them, they cry from their stomachs or their solar plexuses, so I can hear in their sound—welling, vast and dry—both ocean and desert, so I can hear the prayer, their praise not of the thing they are peddling per se but of its cultivation or capture. The voices, scratched and raw, beseech without begging; they need me, and they don't need me: they create a mood of precarious solitude. Rather than raise their voices as in a yell, these street singers call upon their voices. They cry not *of* the fruit or fish they sell, but *to* it. More than an advertisement, the cry is a way for the solitary vendor to keep himself company in the bustling market whose competing noises may or may not hold him. It's more than an announcement, because I hear in its sharp entreaty, its appeal, and its want, the sound of hunger.

~~~~~~~~~

Kolya used to like to run his fingers along the side of a balloon to make a squawking sound, then see how many squawks and pinches it would take until it popped. I used to like rubbing a balloon against the fuzzy surface of a rug, then take pleasure in the way it would stick with crackly static to different body parts. I don't think this has anything to do with gender. If I let Kolya's screaming photograph help me to remember one of my own mood cauls—a zone both sheltering and porous created by sound—I find myself returning to a very early scene from childhood set before the glow of a TV set, a glow that seemed an effect of its shutters-drawn evenly lambent droning. I almost wrote, "in the dream," but it wasn't a dream, it was an episode, and it wasn't an episode on TV but in my life. I was too young at the time to understand images or words, so it's not really possible that I was following whatever was on TV. I was held, however, inside the television's emanating light and the sounds that came through its coarse and fuzzy speaker. I was busy with the unself-conscious activity of cutting my own hair with a pair of blunt-edged pink-handled child-sized scissors. I wasn't trying to accomplish anything, but I was busy feeling something—something that sounded like what I sensed when I lost my feet inside of granules of sand on a beach. I was timelessly exploring, when a scream burst through that was at first my mother crying, "Oh, no, Mary, what have you done!" followed by my own drop out of the zone and imitative screaming in turn.

The number of lessons I was to learn forced me to attention:

(1) one is never to cut one's own hair—that's only something an in-sane person does; (2) scissors are not to be played with; (3) your own head is one thing and your doll's head is another, though you also shouldn't cut the hair off your doll; (4) bubbles are not safe zones; moods are easily broken by the cries of adults. Life breaks in to terminate the mood, leaving me to wonder whether moods have their own temporality, their own terminus based on their own end point. I mean, how long would my hair-cutting activity have gone on inside my bubble if I hadn't been stopped? When do you decide you are finished? When the body calls you back with hunger, a bowel movement, the need to sleep?

How long is the length of your scream? What form does your cry take? Which of your cries has gone unheard, unanswered? When is a mood the effect of a scream that doesn't require the presence of another?

~~~~~~~~

We speak of people creating bubbles around themselves when we want to suggest a self-insulating shield of complacency, immunity, or ignorance. But insofar as we are all equipped with voices capable of generating heat and light, tone and atmosphere, we are all reliant on self-propelling mood spheres to carry us through our days. So much depends on the pliancy or porousness of our bubbles, their suscepti-bility to air currents or tendency to stick. Memoirists of depression describe feeling as though contained and cut off by "invisible but im-penetrable barriers," "glass walls that separate [them] from [them-selves] and life," or encased in "coats and coats of darkness." Some-times we need to be a voice crying in a wilderness, to stand at the edge of being and scream into the universe's wide expanse: the void. But the image that writers of depression draw of their experience suggests a mood born of a circling cry or winding sheet. Not merely an image, it's as though depression *is* the reverberant crash of the cry violently returned to one, surrounding the self or shattering it. In a different state of affairs, a cry is received, reformed, or, floating off, dissolves to make room rather than hem us in; in lieu of a glass cage, each minute, and each day, a dissolve that recedes like bubbles in a wave until the next dissolve appears.

There are bubbles we create with our cries—our moods—and there are bubbles prefabricated and inescapable, the sometimes perverse, even though well-meaning, concoctions of our fellow men, sure to reorient our temperatures, tempos, and states of mind. Consider the "climactic envelope" that is an architect's dream in the design of an entire city encased in a pneumatic bubble. Lisa Heschong, environmental planner and author of *Thermal Delight in Architecture*, explains, with the entire landscape set to the same comfortable temperature (and mood?), the idea of an outdoors is abolished. Technologies of heating and cooling that link numerous buildings or moderate the air in a single building can have a similar effect. Their aim is to achieve "a constant temperature everywhere, and at all times" but "this uniformity is extremely unnatural and therefore requires a great deal of effort, and energy, to maintain." In the developed world, we may have gotten used to and come to expect an evenly moderated indoor climate, but that evenness is far from natural, and not commensurate with thermo-sensory conditions that produce pleasure in us. Heschong offers the example of the delight humans take in exposure to a range of temperatures and in contrasts, or extremes juxtaposed, for example, of sitting by a fire while a storm rages outside or plunging into a cold ocean after baking in the sun. She argues that our nervous systems are more attuned to change than to steady states and proposes a "thermal sense" as its own faculty, "for we have specialized nerve endings whose only function is to tell us if some part of our body is getting cooler or warmer." Thermal nerve endings, she suggests, are "heat-flow sensors not temperature sensors" and are thus never neutral: "They monitor how quickly our bodies are losing or gaining heat." A continual and subliminal sensation of "the heat flow of our bodies," Heschong writes, "creates the general background for all other experience."

Is that mood lurking in the background as an all-defining backdrop, or is it heat? And what of mood, heat, and sound? In a jam session, things get cookin'; a walk in snow is tantamount to the sound of snow beneath my feet, and the mood is bright-calm, calm-bright. None of these conditioners acts independently of the other, per se, but coincidentally, even if we do our best to section and segment them off: to make a dreary neatness of experience.

~~~~~~~~~

Within the first ten seconds of birth, each of us takes our first breath when our central nervous system responds to the sudden change in temperature. It sounds like a gasp. A developing baby produces two times more heat than an adult. A newborn might be covered in a thick waxy substance that helps a baby float in amniotic fluid. It will slough off in the baby's first bath.

When Kolya turned six, I gave him a story that I wrote for him about a cardinal. I can't remember the story's plot except that it was inspired by the suddenness of the bird's red presence in the garden one early April afternoon. I remember bestowing the bird with the power of anointing the boy who found him with just that: a mutually supervening presence of witnessing and being witnessed in turn. When Kolya turned eight, I gave him morning glory seeds since, never expecting to outpace his high-achieving sister, and, unbeknownst to him, greens flourished in the plot his parents had given over to his care. "You have a green thumb," I told him then. "A green what?" he screwed up his face and scratched his head; inspecting his thumb through squinting eyes, he shrugged. "Whatever," he said, and, "Hey! Will these morning glories climb the fence?"

I always gave Kolya books, but as he approached eleven and then twelve, finding the right gift became difficult. He'd moved through a *Minecraft* stage fervently and swiftly, and now his head lay buried in a game he played on his phone in which the physiognomies of actual soccer pros, tiny as cells under a microscope, scrabbled and pitched inside a field Kolya could hold in his hand.

"The best gift you can give a twelve-year-old boy," a friend of mine explained, "is a lava lamp." She was right. He was thrilled.

What's the deal with twenty-first-century boys and lava lamps? Their oscillating bubbles in incandescent hues are mood simulators par excellence. They set a pace of even-keeled ejaculation for a boy on the brink of losing control. They're compressed screams, bottled, now gently gurgling.

Kolya rarely screams now. He's very dear, and very sweet. He's gentlemanly, and it almost breaks my heart the way he still hugs me unabashedly, still turning his ear to face my chest, the older we both get, hello and good-bye. The most recent footage I have of him once again places us in the garden. It involves water in slow motion, and the staged humiliation of being soaked by it. It's August, by the looks of it, when everything's at its blooming best and end. It's hot. Kolya is showing me and Jean and his father how to use an iPhone to make videos in slow motion. First he directs his father to stand at expressionless attention while pulling the trigger of a spray bottle, once, twice, three times for a five-second-long show. The light on this day couldn't be better for capturing the slowly articulated mist and, with it, sounds—coiled and elongated—sheathed by the spray: the camera-phone's on/off button now a xylophone tongue struck by a ball-shaped mallet; bird chirps turned to underwater space creature boings; high-pitched voices hammered into low-lying animal growls. A second video is more elaborately staged but equally short and amusing: father and son face one another in their matching blue T-shirts. Kolya pushes his father whose rehearsed reflex—complete with overblown exasperated facial expressions—is to toss the water from his wine glass onto Kolya as he bounces back. This is the point of the film: to capture water in slow motion as a beaded pelt and splat as it traverses a distance of two feet between two facing bodies. At the end of the water arc, droplets fall to make a necklace of dashes that punctuate Kolya's chest and chin. He's plastered, but beneath shut eyes, he beams a full-toothed smile.

Now the birds are aharmonic, like those in an Amazonian rain forest while a slow ultradeep human voice can almost be heard to say, "Don't ruin the iPhone." This, no doubt, in reference to the liquid nature of the scene. The rose petals, singed at their edges, emit a pom-pom drumming, their scent made more sugary by the damp. Red geranium petals peal, and the dogwood releases its coppery downpour in the shade. The point of the game is to make something mind-bending and, in the process, to get drenched. Like his four-year-old scream, the gale is tuneless and unstoppable. It creates a mood of madcap inside an urban paradise. For five full seconds, we all laugh and laugh and laugh.

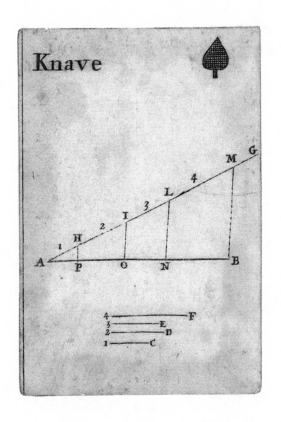

(charts)

I can imagine science being used as therapy for all sorts of alienation-related psychological disorders. What better way to bring an errant mind back into the fold than to give the patient a stopwatch, a ball, and an inclined plane and tell her she can have lunch as soon as she comes up with something interesting to report?—Barbara Ehrenreich, *Living with a Wild God*

But how can we uncover atmospheres and moods, retrace them and understand them? Is there such a thing as a professional—or, for that matter, "scientific" approach?—Hans Ulrich Gumbrecht, *Atmosphere, Mood, Stimmung*

is it possible to die
of a feeling?

Is it possible to die of a feeling? And how might that be different from dying from a grief one never recovers from, or those tales of people who die at the same time or very near to the person they lived with for decades of their lives?

I'm not talking about dying of depression, per se, or an inconsolable sorrow. I'm not even talking of dying of love, unrequited, or just as passionately, fulfilled. I'm thinking of a fear I've maybe harbored all my life: that certain members of my family might—what might they?—but here I have to pause to reckon with the strangeness of it: die of a feeling. It wasn't a "spontaneous overflow of emotion" as prelude to a Wordsworthian art that they were prone to, but the prospect, if they weren't careful, of being carried away by emotion to a place of no return. You'd never hear the phrase, "he's ready to explode"; in my family, people *would* explode. Some families are of the crock-pot or slow-cooker variety whose feelings simmer and brew; others are of the carbonated beverage type—"bottled up," they're cantankerous with too much gas. Mine was the sort unable to keep feeling under wraps; blood boiled, it bubbled over, a "gasket" blew as tribute to pure and simple rage, or a vessel threatened to shatter and break when a voice sounded too high a note of over-elation. Could the fierceness of a crying jag make a person revert entirely to liquid? Cause of death would be cut and dried, for there was no room for subtlety amid these credenza-sized feelings: the death certificate would read, "fit of tears occasioning loss of all bodily fluids. The subject dried up."

I once did see a person die of a feeling—fear—but it was in a fairy tale on a stage in a jewel-box-like musical theater, the Bolshoi. It was a few weeks after 9/11, and a troupe was performing *Pikovaya Dama*. A shock of white hair that was unmistakably that of stroke-recovered Boris Yeltsin shone from a balcony where he raised a stiff hand in greeting, but this was just a prelude to all that would conspire inside a color wheel of burgundy, gold, and the black dress of a dancer studded in diamonds that flickered as she shook. She didn't just up and die of fear or flat-out faint from an upswell of panic. Every molecule of her being quavered—even at the distance of a speck from where we sat, each and every part of her being articulated tremor until she filled the stage with it, and then the theater, until she slowly, like a fire pelted with waves of steadily intermittent rain, went out.

I could say the dancer was *embodying* a feeling, whereas my family was always *expressing* them; that she made her body into a living *emblem* of fear, whereas my family's bodies were feeling's ill-equipped *vehicles*. Our moods are the residues of familial feeling. That's one way of putting it. Such familial feelings, and the means by which they are conveyed, create a milieu—in my case, of precariousness and heightened intensities, bold hues, and dark swirlings. Which isn't to say I played the role of a quiet figure in the corner observing my family's pyrotechnics. Think of me as the confetti-torn product of its bang and begin to conjure your own palette of tones born of the worlds that formed you.

~~~~~~

I'm not writing in a Poe-esque fever dream, though I am a little shaky, adrenalized on too-strong tea, and ideas. I'm writing in a muggy attic room—dispel any thoughts of Baudelaire!—with a fan running. It's the type that turns from side to side surveying the space with the occasional reminder of a breeze. Afternoon sirens "scree-ing" in my city neighborhood; the sleeping form of my cat, Bug (a major player in these pages); the squat black anvil of a landline that no longer rings; bricks mortared by nineteenth-century fingers; the shadow thrown by an umbrella in black and white pushed open on the beach of an ancestor's Adriatic according to one photograph I keep; a display of dice in Lucite—yellow, red, and green; and, the sense of my knowing where everything is amid bolts of piling paper that a different set of eyes might retreat from as "holy mess" work in concert to bear the notes I'm trying to create, for what is that prac-tice that defines a life—writing—if not the articulation of a longing for a particular tone or voice or Mood?

As a baby, I hold onto an ear; as a toddler, the handle of a tea-cup; as a preteen, the hard-bottomed back of a book. Coalescences comfort me, so I condense all memory of my parents to two pieces of clothing: a brown wool jumper; a cleated blunt-edged work boot in black.

What does any of this tell us? If the enunciated was always over-taking the ineffable in my family, was the possibility of mood erad-icated in the process, or did that lay the ground of its necessity? I know our household was equivalent *to* my mother's moods (see the walls at one time painted tangerine orange and cobalt blue), while my father's less articulate body of feeling grumbled below ground like a snoring man or a volcano that at any time might erupt. Oh, but moods distinct from feelings can be hard to parse.

Less than a decade ago, a group of British psychologists admitted what everyone in their field already knew but maybe weren't willing to say: that the criteria used to distinguish moods and emotions vary widely from "physiological and neurological through to behavioral and social" and that most of the time the distinctions are arbitrary and unfounded. Even the most fluent of writer-neurologists, Anto-nio Damasio, admits the impossibility of finding a suitable model or a set of terms. Some emotions burst; others follow wavelike pat-terns; some emotions count as background emotions; others are pri-mary and secondary in his scheme. Variations on these are boundless depending on "circumstances and individuals," but when an emotion becomes frequent or continuous over a long period of time, it seems to come to count as a mood. If you sound a particular emotional tune frequently or change your emotional tune often, you are moody. Moods can be pathological and then we speak of mood disorders. One can see why Damasio relegated this review of the state of the art of our understanding to a footnote.

The British group, noting the limited progress of traditional empirical approaches to the subject, agreed to turn to what they call "folk psychology." Let the people decide. People must know what it is to have a feeling, over and against a mood, and if we're claiming to treat both, shouldn't we know what we're treating? We're not asking them to offer opinions on abstract psychological problems like the nature of consciousness, they said, but to offer some insight on something they might have intimate knowledge of. The results were not always congruent, and far from conclusive, and doesn't so much of this come down to how we are culturally tutored to divide and conquer ourselves? Bifurcating parcels abound in the answers participants supplied, as well as some surprising suppositions: emotions are public; moods are personal. Emotions have positive connotations; moods have negative ones. Moods have to do with thinking; emotions have to do with feeling. Emotions have specific causes; moods do not. Duration is a standard figure for differentiation, so we get: feelings are short; moods are long.

"Language does not always represent psychological reality," the authors warn, but as a writer, a student of language, I've come to believe that language will supply not just the answers but the questions we have yet to pose so long as we care to study it, trouble it, work with it and let it do its work, and play, in turn. We just have to be OK with the matter language yields being as multiform and beguiling as the questions made possible by its terms.

Feelings are things that our language says we "have"; moods are things we are in, and they lack a verb form: we never say we're "mooding." Feelings can be transitive and intransitive: sometimes

I'm feeling *something*; other times I'm just good old feeling. We claim that it is possible to be periodically devoid of emotion, but we're never without mood, and the evacuation of feeling could be the premise for a mood. Trying to make feelings line up with moods, on the other hand, is futile: in a bad mood, I can feel irritable and excited, whereas a good mood could conjure the feeling of hunger or lack.

Maybe the whole point of feelings and moods is to test the limits of our imaginations, requiring a moment-to-moment re-creation of images that gambol and lurch, elegant as monsters caught inside the scrim of a shadow-puppet board. In the land of metaphor, distinctions only serve to create a bridge for new relations. Try: mood is the sea, the tide of our being; feeling, the boats, buoys, and traps that sink, float, skim, rise to surface, require a little dredging, dissolve, erode, or anchor in. Feelings are lit from above on a sea of mood, and propelled from below. As such, they peak or churn.

Returning to the familial hand that I was dealt, there was a King of an impossible mood, a Queen of an exuberant one, and a Jack of intractable depression. And where did I fall in the deck? My moods are even-keeled to nonexistent. Or so I misperceive them, and thus this book? A lot of the time, I think I'm in an interested or curious mood, and writing, which I do a lot of the time, puts me in a strange one.

I suffer a recurring mood endemic to the American middle class, or perhaps it's a mood of unease bred of Catholicism: a feeling that something's not right or something is *just* not right experienced mostly in moments when "everything is right." I'm intrigued by this mood for its being a paradox. There's no feeling I can't turn into an intellectual problem, and that's what saves me from being embodied. Embodiment really isn't all it's cracked up to be, and I'd rather have nothing to do with its flab and hunch and pain. Bodies fail to mirror us—I don't know anyone who feels at home in theirs—but they are said to mimic our moods, so there must be something to be said for mood as physiology, even if we conclude that moods aren't exactly to be found *in* bodies but are more like amniotic sacks, the body's ghostly Other and originating twin.

~~~~~~

"My father was a proctologist; my mother was an abstract artist. That's how I see the world," was a line I always loved from one of Sandra Bernhard's comedy skits. If I follow the scent of the psychological literature, I find that my parents fall pretty squarely into one of two camps: my mother was emotional, and my father was moody. The article by the Brits summarizes the distinction thus: "A moody person is one whose reactions to an event are likely to be more consistent with the (usually negative) nature of his or her mood than with the actual nature of the event"—prone to bad temper or depression; whereas an "emotional person is one who 'often reacts in a manner consistent with the nature of the immediate event or situation, whether positive or negative' . . . but in an intense even extreme way; someone who perhaps cries while watching a sad film, or who is easily and demonstrably angered by minor irritations." Everyone has emotions, but some people are emotion-*al*; everyone has moods, but some people are mood-*y*. Those three little letters exert an awful lot of interpretive pressure on individual humans, and I'm not sure I'm ready to cash in the complex personalities of each of my parents for an adjectival flourish. Maybe the point is that some personalities draw our attention to their emotions while others ask us to pay more heed to their biceps; some want us to be aware of and wary of their mood-creating capacity, and others want us to think of them as very good cooks. This much could be true: where my mother was prone to be carried away by emotion, my father made a dwelling of mood and remained stuck inside its monochromatic walls for the remainder of his days.

Formulas only sort of work; they are fine for the instant, but not in the long term. My mother used to recount a two-part recipe she relied on in deciding to marry my father: (1) unlike all of the other men she dated, he didn't try to rape her on the first date; (2) when her two-and-a-half-year-old niece died of spinal meningitis, and she told him what had happened, they sat together in his car and cried. Not only was he not the sexually aggressive sort, but he was a man who cried. She thought from this she had found a different *type of man*. But she was wrong.

How often my father bulldozed the walls of our house right down to the ground with his rage, then left my brothers and me to pick through the rubble in search of stray popcorn kernels, a crushed pair of spectacles, a fossil made of the haywire that was once cotton candy. This is how I picture us: set before a scene of expectancy, the lights set to low, hardly able to hush our squeals before a puppet show proscenium of rising curtains when the wrecking ball that was our DAD would blow in. Already tiny, we became tinier, like birds in the aftermath, hopping from jagged peak to jagged peak. We didn't die. We pecked. We flew, even if the dust had long since settled.

the flower inclines
toward blue

What is this shadowy creature that won't allow us to take aim? Mood. Yet, in the 1970s United States, we thought we could wear and register our moods on our fingers if not our sleeves. The same era that bore the "mood ring" bred Tompkins and Bird's *Secret Life of Plants*, a book that took "flower power" literally, pursuing the possibility that plants were sentient, "conscious," psychic, and capable of communicating with us humans. It was the same era in which some Americans put their minds to devastating whole populations of people in Southeast Asia with chemical weapons, and the epoch that saw Maude of sitcom fame put the phrase "manic depression" into public discourse in an episode in which she zanily proposes that Henry Fonda run for president. The writers of the episode, called "Maude's Mood," have a doctor figure submit to its American audience as a matter of fact that Maude's intense enthusiasms and dramatic lows, her "manic depression," is a biochemical disorder treatable by drugs. Come to think of it, Jimi Hendrix had already assimilated the diagnostic category to his art in 1968 with "Manic Depression," a song made of fast-moving upscale and downscale guitar glissandos and percussive snares whose lyrics move from the glibly moody, "manic depression's a frustrating mess," to the subtler Mississippi blues wish to "caress," "music, sweet music" as antidote to a downward spiral.

Amid a contemporaneous lexicon of turn-ons and turnoffs, Hendrix's "Manic Depression" says, "Well, I think I'll go turn myself off and go on down" not to the Mississippi but "all the way down" because "really ain't no use in me hanging around in your kinda scene." Who can reconcile the era's scene at a time when individual people were understood to be part of their own scenes—the self-made milieu that matched the parameters of one's *pad* whose colors it shared—with *the* scene, the infinite scene. I think of the late '60s and early '70s in terms of an antinomy: a languid psychedelia of bodies falling backwards into fields of white and yellow flowers at home and into scorched and poisonous red and orange rivers of death and deformation abroad.

Like most twelve-year-old American girls in 1972, I owned a mood ring. Rather than protrude from its setting, my mood ring's jewel was flush with its band, making it seem all the more futuristically slick, seamless as *Star Trek*'s prescient spandex, and just as extraterrestrial. Made of thermo-chromatic liquid crystals that emit a change in color based on change in temperature, the mood ring exerted a fascination, in spite of what *Wikipedia* has to say about it in hindsight: "No direct correspondence between a particular mood and a specific color has ever been substantiated." I don't think anyone judged her mood ring for faithfulness or lack thereof; irrespective of their fidelity in *reflecting* a mood, mood rings *created* a mood and were part of a scene.

I remember the mood as one of clamor as people clustered around each other's mood rings, as around a Ouija board, to see what it might tell. I recall its kinship with other interpretive models that I was preoccupied with at the time: in particular, handwriting analysis, physiognomy (it may as well have been 1850 rather than 1970), and palmistry. What could I have been after in afternoons spent inside library stacks where, instead of reading, I traced the handwriting of my teachers, friends, and family, having struck upon a discovery that my handwriting analysis book did not pause to yield: I remember the feeling of a quasipleasurable if startling alien invasion, like the sudden assumption of a second skin—with what intimate, immediate, pulsing, body-full force one could experience being someone other than oneself by tracing the handwriting of another person. Feeling another person's handwriting carry my pen along, I'd give myself over to a stranger's speed or care, loop or flourish, hollow or hunch, magnitude or spirit, float or press, carriage, dance, crib . . . mood?

Why do attempts to find oneself in forms where one is not so often engender the need for dominance? Sad to say, if one week I was letting myself experience what it felt like to be disarmed, the next week I was keen to develop the skills of a forger—swindler or mountebank could not be that far off. I was learning to imitate that most particular of handwritten impressions: the signature.

The mood ring was part of this surround. I'm not so sure I sought a portrait of myself in its telltale palette, or if its appeal was its implicit though unarticulated offer to enable one to feel what it was like to be someone other than oneself—in this case, to try on a popular idea of personhood, a person with passions ranging from gray to bright blue, nearly purple, a person who could don the equally loud-colored clothes of a largely loud and increasingly superficial age (the '70s, after all, were a prelude to the '80s). My mood ring drew a sacred circle round a local scene: it did its best and most frequent work on the three front steps that characterized all of the homes in our working-class neighborhood, as if steps in front of hovels could make one feel one's meager earnings still took oneself and one's family "up" in the world. I recollect it there on the three front steps where I sat with my friend, Patty, our heads hovering over its small pod of color like a secret well into which we peered on one summer afternoon after another of "nothing to do."

The mood ring was part of a mood, created a mood, and conjures a mood to this day: of my manic aunt and my depressed mother, or my depressed aunt and my manic mother, singing while they washed the dishes following a large family meal at their parents', my grandparents', house. Together, they belted out each verse and line, while they also harmonized, "Gonna take a sentimental journey, gonna set my heart at ease." The volume and intensity of their voices would heighten when they got to the wonderful effect of the forced rhyme,

"never thought my heart could be so yearny," until no one could say
that dishwashing was something you couldn't put your heart into;
as if to accept without accepting the drudgery of their gender lot; as
if to say there would be no suffering in silence here; as if to match
the depths of their unsatisfying marriages with the hilarious bliss of
their being in on a secret; of their being in cahoots with rage.

What I remember of my aunt, my dear aunt, was how, at one
such family gathering as this, my father noted that she was puffy
and thought she should see a doctor, and how near to death she was
days later until she died long before her time, swiftly and without
explanation, from kidney failure, though some said the real cause
was depression. Since I'd never heard of death by depression outside of suicide, I wondered if they'd meant that she died of sadness.
Did they mean she died from none of us having acknowledged her
depression? I only knew her as a person bursting at the seams. And
how she seemed always to be gargling her words as they bubbled up
uncontrollably. And how I would find myself watching when nobody
knew it the way her veins pulsed in her ever-twitching ankles. Quivering, restless, how like her name, Frances, was to her state: frantic,
frenetic. I remember my aunt Frances for how she smothered me
with hugs and gave gifts on days other than birthdays: one day, a
figurine she said reminded her of me, that put me in her mind. I
remember being so glad that she had noticed me, even though I had
always felt her love, and I pictured her in the gift shop alone with the
idea of me. I remember how she filled me with a feeling that must
be the opposite of depression. The little gift said that Aunt Frances
thought of me when I wasn't there, and even on days when it wasn't
my day to be remembered, and how on one such day she gave me
a statuette of a girl with her mouth open, singing, while playing a
mandolin (fig. 7).

~~~~~~~~~

Mood rings aren't generally readily available these days so when
I saw that they were offering them as party favor type prizes at a

(7. *I wondered if they'd meant that she died of sadness*)

hipster-doofus circus outside Vancouver B.C.'s retro tiki bar, the Waldorf, I asked if I could buy one. The ring leader who manned the prize booth proposed ninety-nine cents as a price for the ring—a plastic job with a large, oval stone—and I accepted, delighted to discover that the price also included a perfectly square miniature sheet of paper with a code. "Mood rings allow you to see your true emotion through color changes in the stone! Your own energy will cause the stone to change depending on how you feel," the cheat sheet explained, and pictured alongside bands of primary colors the following explanatory grid:

> Blue: happy;
> Light blue: relaxed;
> Green: active;
> Yellow: alert;
> Pink: romantic;
> Red: excited;
> Black: stressed.

I no longer had my mood ring grid-of-yore to hand, but in an Internet search, I found one that claimed to be vintage that didn't quite line up in either color or meaning with this newer model. It read:

> Black: frigid;
> Gray: irritable;
> Yellow: melancholy;
> Green: cuddly;
> Blue green: amorous;
> Blue: sensuous;
> Dark blue: passionate.

Whether this version is the same that I was given as a preteen is no matter. Alignment was never the mood ring's keynote, as this chart reminds us: the mood ring has probably always been intended as an implicit or explicit sexometer of crude proportions. What's most pleasurable about the mood ring is our being in on the secret of its clown-faced farce: because we know, don't we, that homologies don't really hold between the states of passion, warmth, and happiness in human beings; and we know that old adage that a cold hand might hide a warm heart; and we know that there are better ways to spark or end a love affair than with a mood ring. Still, the mood ring teaches something about mood if we consider that its crystal stone not only registers the temperature of its wearer translated into light but also reflects the immediate colors that comprise her external world. It's this combination that the mood ring yields without admitting.

The trouble I always had with my mood ring was that its mesmerizing phosphoresce of colors never matched the monochromatic color of its code. It could be streaked with dark in bursts of violet or centrally pink with yellow-green margins. Not blue, but cobalt, now ultramarine: the colors in my crayon box had more amplitude than the colors offered for my moods. If the colors failed to match, their correspondences were also worthless. What code could supply meaning to something ever out of view? How could science, pseudo- or otherwise, evidence the underside of mood: some impossibly beautiful variance, and chance?

In a bracing cultural analysis of what was once called manic depression and is now called bipolar disorder, feminist sociologist Emily Martin, who has herself benefited from pharmacological treatment of the condition, details our penchant for charts and their irresistibility in studying mood. *Bipolar Expeditions*, Martin's book on the subject, considers the treatment of mood disorders as part of a nationalist project to create a properly capitalist citizen complete with optimally manipulable mood. Mania isn't to be eradicated; in fact, it is regularly encouraged in everyone from the most monumentally successful business leaders and politicians to creative geniuses, artists, and Hollywood producers, directors, and actors, New York cosmopolites, and the consumerist man in the street. "In the US economic system, there is a premium on measuring and tracking any valuable resource, and that includes moods," Martin writes, reminding us of the relationship that inheres between tracking moods and gauging consumer confidence, or arousing what John Maynard Keynes famously referred to as the market's reliance on people's "animal spirits."

Even the simplest of mood charts is not as transparent as it seems. For one thing, as Martin points out, today's mood charts are micromanagement systems, unlike those designed by Emil Kraepelin, the nineteenth-century eugenicist classifier whose diagnostic nosology is the basis of today's largely biology-driven psychiatry and its *Diagnostic and Statistical Manual*, editions 1–5. Kraepelin's charts, though telescopic, may have anticipated the microscopic methods of today: "In Kraepelin's case, an individual's entire life span could be described on one page; today, one page usually contains the details of only a single day. Each day, in turn, can be divided into periods of hours and minutes and each quality or activity can be registered practically by the minute." Subsequently, today's charts are distinguished by their reserving a significant space for recording "the means of ameliorating mood disorders"—in other words, they are companion accounting systems not only for establishing norms of mental "hygiene" but also for titrating the distribution and dosage of a vast array of pharmaceuticals aimed at enhancing or modulating their subscriber's presumed mood.

Does a doctor keep a record or chart of or for the patient, or does the patient maintain his own map? Are the records interpretable by one, the other, or both? Does a "normal" person's mood state fail even to register on certain charts, implying that only a person to whom mental illness has been attributed *has* a mood? How is lucidity registered and aberrance pictured? Is the ability, desire, or need to make a chart indicative of rationality, prelude to scientific discovery, and bedrock of medical practice, or does it resemble the irrationality of obsessive-compulsive disorder? These are some of the questions raised by *Bipolar Expeditions*, to which I feel compelled to add: what category is reserved for folks like me who are "chart averse"?

I am incapable of keeping a chart of or on anything let alone moods, but I can be meticulous, fastidious, dogged, and highly focused, and I have an eye for lists. In the literature on the psychology of mood, lists are bounteous. There are lists of mood-measuring tools based on statistical methods hailing from the 1960s and 1970s, the MACL (Mood Adjective Checklist); the POMS (Profiles of Mood States); and the 8SQ (Eight State Questionnaire). There are lists of pioneers who set their mind to mood or to distinguishing mood from emotion: Vincent Nowlis (the behaviorist who developed the MACL); Alice Isen, a cognitively oriented social psychologist; Edith Jacobson, a Freudian psycho-dynamacist; Karl Pribram, a neuropsychologist. Nowlis may have been onto something with his potentially nuanced checklist—aggression, anxiety, surgency, elation, concentration, fatigue, social affection, sadness, skepticism, egotism, rigor, and nonchalance—but researchers attempting to apply his categories inevitably fell back on summarily dyadic terms by which they grouped them: the ever-popular "positive" and "negative."

The list of monoamines that work as neurotransmitters most associated with mood in our neurobiologically minded contemporary age feels especially sinister to me—dopamine, serotonin, and norepinephrine—since by now we all recognize their accompanying, most widely prescribed, brand-name drugs, in singular or as cocktail, not Mallow Cup, Butterfinger, Milk Dud, Charleston Chew, Milky Way, or Almond Joy ("sometimes you feel like a nut; sometimes you don't"), but Ambien, Ativan, Depakote, Effexor, Eskalith, Klonopin, Paxil, Prozac, Risperdal, Wellbrutrin, Zoloft.

Sometimes the lights of psychological observation seem set to low—what a friend of mine means when he applies the word "dim" to people's unwillingness or inability intellectually to engage, but this doesn't mean that the seeming simplicity of certain mood manipulation studies fails to delight or shed light. One has to appreciate what feels like a sophisticate's innocence (or idiot savant quality) to Alice Isen's mood manipulation experiments in which she sets out to prove that physicians who receive an unexpected bag of candy make better medical decisions, or when she observes similar lifts of mood and the tendency to be generous or helpful when people find a dime in a pay phone, or are given a cookie while studying, or delivered free samples while shopping in a mall. It's a no-brainer to conclude from such studies that people's moods can be manipulated or easily changed; what's more enchanting to me is how often sweets figure in this work as motivational treat.

Layfolk might find themselves asking if we really need science to prove some rather self-evident propositions, in the same way, without benefit of art historical instruction, people feel deceived before abstract art, claiming their two-year-old could do the same, but better. "Who pays for this shit!" my partner, Jean, demands to know while I read to her a *Huffington Post* headline for April 15, 2013: "Hedonometer Happiness Index to Chart Moods around the World." "A stock market-style 'happiness index' that measures the mood of the world has been launched by US scientists, charting the mood of the planet on a graph that rises or falls, based on tweets," the *Huffington Post* reports. If the presumption of charting the mood of the planet is preposterous, the "methods" are egregious: let's just say, our desire to collect information and interpret it reaches all-time absurdist lows with "data mining."

In this funded initiative linked to a mathematician at the University of Vermont, "paid volunteers" are asked to rate the "'emotional temperature'" of words of "some 50 million tweets from around the world" daily. Assigning words "happy," "sad," or "neutral" ratings based on ridiculous one-to-one correspondences of words (e.g., "haha" rates a "happy"), the team hopes to create a "minute-by-minute barometer of global happiness." Their conclusions are as staggering as their scoring system is sophisticated: their online sensing device, or "hedonometer," showed a drop in global happiness on the day of that year's bloodied Boston Marathon. They also determined that Napa, in the heart of California's wine-growing region, is "the happiest city in the US." (What's the difference, anyway, between wine and candy?)

An (unpaid) blogger who created his own visualizing device with a similar aim but based on "positive or negative emoticons" recorded in Tweets across the world also hoped to create what he called "The Man in Blue Mood Map" of the world. "The Man in Blue" reaches somewhat more subtle conclusions: that there are no patterns he can discern on a global scale, and that, even when full-scale disaster befalls a large population of humans, "everyone's pretty wrapped up in themselves."

Even if the *Huffington Post* article is meant to be read tongue-in-cheek, the journalistically tonal giggle—let's call this a twenty-first-century happy mood of purported objectivity—can't negate the seriousness of the mathematician who claims, "we're measuring something important and interesting and . . . sharing it with the world." The certainty of measuring mood follows from the mathematician's disclaimer that he and his team aren't trying to "to define the word ['happiness']." But don't they have to define it in order to measure it? And how can their interns rate words for their happiness quotient without "defining" happiness? Worse yet: where is a model of mind here to match a theory of language: words are not bearers of meaning in themselves but depend on convention, on context, on history of use and individual inflection; nor do words enjoy a one-to-one correspondence with feeling, or mood: the use of a happy face emoticon or of the word "happy" in a daily tweet—a highly conventional mode of conveyance in itself—might be entirely arbitrary in its relation to one's mood state. And how about all those people who do not care to tweet, and whose presumed moods are not part of this measurement: are they not part of "the world"?

Psychologists acknowledge that "mood" enjoys a "history of casual use"; that the source of our moods is obscure; and that "lack of specificity is [mood's] most defining characteristic." Yet we continue to measure mood as a way of knowing it, communicating about it, and medically treating it. Mood may thwart its own evidencing—or, at least, resist our crude epistemological tools.

The effect of our charting and drugging that which we know so little about might be to eradicate madness, according to Emily Martin—to quell unruliness—or eventually to replace madness as we currently conceptualize it with a new notion of madness yet to be described or socially deployed. Martin draws upon French philosopher Michel Foucault's formulations in prophesying the disappearance of madness as we know it, predicated upon two conditions: "First, precise pharmacological control of all mental symptoms, and second, rigorous definition of behavioral deviations accompanied by methods of neutralization."

Neutralization as a form of stifling applied to enemies in real world contexts of war, or, in sci-fi contexts, to alien creatures with stun guns, is always a cover for some other form of more extreme violence that goes unspoken. When I read in the psychiatric literature that research is stymied when clinicians have to reckon with the fact that an individual may be in a "positive and negative mood at the same time," I experience a twinge of self-neutralization: as though psychologists have made themselves dumb and numb in order to continue to measure what they hope to control and modify. They have lost their appetite for conundra.

This isn't always and ever the case, and it's possible to get excited all over again by tuning into theories like the ones that William N. Morris summarizes in *Mood: A Frame of Mind* when moods are distinguished from emotions for being generally informational in nature—moods as barometers, monitors, or cueing mechanisms in a "self-regulating system"; or when moods are understood as residues of receding emotion, their more mercurial aftereffect; or when the idea of focal attention is brought to bear on mood disorders: I've wondered, if we all are subject to our moods, is a person suffering from a "mood disorder" one who is more conscious of, acutely aware of, or distracted by her mood state? In other words, is a mood disorder a disease, or a state, of *attention*? And what of rhythm? That we are creatures dependent on sleep-wake oscillations, cyclic intervals, synced or not, to the presence of light, minutes, or hours, or days and nights—our circadian and ultradian rhythms—might play a part in a relative absence or presence of a feeling of "mood."

What continues to astonish me is that we claim to treat mood (presuming it's an it) without really knowing what "it" is. With particular reference to the ever authoritative *Diagnostic and Statistical Manual*, Emily Martin reminds us that, "what the terms mean, how they should be applied, and even whether doctor or patient will get to apply them are all matters of contention." Experiments, tests, charts, questionnaires, and mood rings, too, don't really measure mood; they produce an idea of it that is possible to manage. "In the process of coming into being," Martin argues, "the public health crisis in moods has changed the way people experience their moods. Increasingly, there seems to be one universal set of mood categories that everyone experiences. Through the simple act of recording their moods in terms of these categories, people form the habit of thinking in terms of a standardized taxonomy of mood."

~~~~~~~~~~

It wouldn't take much to conjure an alternative lexicon of mood or even to play more attentively upon the harp strings of the words we daily use to give form to our everyday or aberrant states of mind. Starting with "mood modulation": mightn't the phrase be redundant, and what do we do with that? Or, think about the way the word "depressing" figures when my friends and I concur that shopping at a place like Job Lot or Building 19 is "so depressing." Do warehouse-like stores fail properly to exert a consumer coddle? Or do the masses of mounds of discontinued or unpatented stock in such bargain outlets lay bare the vacuity that regular shopping also fails to fill? Are we "depressed" because we're loathe to confront the lengths we'll go to troll for trash?—a malfunctioning beanie copter; a rake whose poor design leaves the handle to break off in your hand with each pull through a pile of leaves; hair products whose contents have separated, making you aware of how unaware you are of the makeup of cleansers you apply to your body and dispense back into the earth. Worse yet, marked-down bottles of what you liked to think of as fancy olive oil—the degree of the extent to which you flush your hard-earned dollars down into the same plumbing system as the toiletries every time you shop: depression sets in when you realize how regularly you are baited by brand names, how daily you are had. Or maybe what "depresses" most in such places is the thought that one of your books could end up there in the 99¢ bin, and with that, the realization of what you try to avoid: that it's already for sale in all manner of 99¢ Internet warehouse bins; or maybe more depressing is that Tolstoy is tossed inside a salad of ragtag dollar-a-dozen self-help titles, here today, gone tomorrow. That's one commonplace way in which "depression" takes on a pasty glow—not just a down-in-the-dumps darkness: it's its shimmer that horrifies; it's a glower, as though to shop in Job Lot or Building 19 is to assume identification with those white-gray figurines so popular in the '70s that caricatured humankind (their prototype being a bulbous-nosed drunk), each of whom was equipped with an epigrammatic joke etched in red to heighten the grotesquerie (fig. 8); it's depressing to think of who bought them—you're certain you even bought one once in the same week you bought your mood ring—with no clue as to the purpose they were meant to serve: as stand-ins for trophies, they were a kind of self-mocking bric-a-brac or kitsch for cash.

Charting things doesn't make me manic so much as it gives me what is called in Yiddish *shpilkes*—literally, "pins and needles up your ass"; a need to move or run, an inability to keep still. When I try to apply my hand to a chart, I feel overcome by a fuzzy, hazy palsy until my crayon falls from my hands and fingers and I begin dreamily to dither or zone out: I will never be able "to stay inside the lines" as the teacher instructs, so why try? Sloppy at core, I will never be able to make anything beautiful. A friend in her thirties took to coloring to quell panic attacks, and I wonder if this meant she was meeting them at their source.

When we speak of elevating moods or mood elevators, a first impulse might be to picture platform shoes (enter the 1970s again). But elevation—apart from oxygen-deprived altitudes—figures for me most vividly, psychically, in dreams in which elevators are always also attached to subway systems (perhaps this is because, in Philadelphia, the nearest city to where I grew up, the elevated transport is called the El). My elevator dreams usually approximate the shape of a letter *L* that eventually gains a leg and morphs into a Z. Elevation is never about levels; it's about sideways movement. Because

(8. *"mood" spelled backwards is "doom"*)

what happens in such dreams is that I step into an elevator vaster
in size than any real elevator—or the size reserved for art or natural
history museums ample enough to accommodate oversized abstract
paintings and dinosaurs. There's usually confusion about what floor
I'm headed toward or misinformation about my destination—the elevator only goes up, but I need to go down; the elevator only stops
on the seventh floor, so I'm forced to wander a labyrinth of hallways
to find the thirteenth. Invariably, the elevator becomes uniquely an

elevator of mind since I know of no such elevator in waking life, as it begins to move sideways, less a ski lift than a sidecar with its own set of unpredictably winding subway tracks.

Direction has ever been at stake in the psychological language of mood, but we're too busy working out the calculus of the relationship between the ingestion of Jujubes and good will toward men to care. One way of figuring mood is to rely on a metaphor of inclination. The words "disposition" and "predisposition" are especially vexing in this regard. Initially an order or arrangement of parts, "disposition" has come to refer, favorably or unfavorably, to a person's nature, constitution, turn of mind, temperament, or mood. In this sense, "mood" is another word for the way a person is put together and, consequently, the way a person tilts. Even though "disposition" has a built-in connotation of a preordained or natural tendency, a person can also have this strange thing called a "predisposition," which might just be a way of saying that a disposition is often enough a *predictor* of how a person will, ahead of time, react to particular persons, places, life events, or things. Not just an inborn or inbred determinant of action, mood in the form of disposition pictures persons as inclined planes when we say that a person has a certain leaning, slant, bent, or sway. We are encouraged to be normal but only really exist as personalities, moods, and minds to the extent that we deviate from a ramrod position: to incline is to deviate *and* to exist: to have a handwriting that is yours and yours alone.

Imagine a mood chart whose scale relies on a spectrum of half-inclined, quasi-inclined, un-inclined, or well-inclined. Then consider that all moods are but approximate: mood as a growing towards not trying to catch up but, merely, in its movement, hinting at a letting through or blocking off of light as when we say "the flower inclines toward blue" and thus imagine its leaning as a process of becoming. Charting color in the psychiatric literature makes for a dystopic loop, from mood's standard definition in the *Diagnostic and Statistical Manual*, 4th edition—"a pervasive and sustained emotion that colors the perception of the world," to the mood ringlike mentality that marketers use in dying mood drugs particular colors: "You never use red for psychotropic drugs," Emily Martin quotes a pharmaceutical executive as saying. "It is said to be bad for psychological or psychiatric problems, signifying danger. Black and gray mean death. Sometimes blue is bad because it can mean poison and can be seen as cold. But then again, light blue or green can be good when you want calm, soothing colors." In lieu of these coordinates, I'd like a technique for remembering the inclinations of my days—to recall the moods of all my living in terms of distillates: a patch of blue sneaker, or yellow mustard stain, the blackberry blot, or blue-black toe, the opalescent flicker in the goldfish bowl, the crushed red of the matchstick's tip, the marbleized rock, the sandstone necklace, the gray menagerie, the loamy soil underfoot, the cinnamon-colored pine strands, the chalky cinders, the onionskin lamp, the black ink on green-lined paper. This same reinspiration of mood could be brought to my present-day perceiving if I were to come at encounters by asking where's the green in this? Where is the yellow-gray? Where's the off-chart gradation capable of altering the uniformity of the display? We could ask this of any situation, not just our hearts.

All of us have different methods, that's for sure, for coping with bad moods. In *Mood: A Frame of Mind*, William N. Morris's helpful review of so much of the psychiatric literature on mood, we read that "survey data indicate that when people are asked how they deal with bad moods, they readily nominate a bewildering array of different techniques." The problem is that Morris and his colleagues don't generally trust the "accuracy" of what is called in the social sciences self-reporting. When they use their own tools to measure coping mechanisms, they come up with four different types of activity as ways of managing mood: "self-reward, the use of alcohol, distraction, and the management of expressive behavior" (by which is meant, body language and facial expressions). Robert E. Thayer, in *The Biopsychology of Mood and Arousal*, prefers the tried-and-true method of a "short brisk walk" (somehow the word, "brisk," reminds me of a brisk cup of tea or coffee and my grandmother's conviction that there was nothing better than drinking a cup of hot tea in hundred-degree weather. By sweating to the point of fever, you would be cooled down by the tea, she claimed).

One day, my six-year-old niece, Sophie, confides, without my asking, what she does when "I don't feel good." "Wanna know what I do when I don't feel good?" she asks me. "Sure," I say, "what do you do?" She tells me she has a secret place she goes to that no one can know about, not even her mother and father, though she recently allowed her best friend into this place when she wasn't feeling well. Now she says she has to find a new place for herself—solitude and secrecy being requisite to a better mood: "Now I go under my desk," she says. Together and alone, alone and together seemed to make a braid of mood modulation for my niece. Away from home as we were, it was hard to find the security of a regularly secret place, and I wondered if her announcement of her bad mood/good mood/go brood spaces was the next best thing to finding her way into one.

I have a friend for whom a bad mood is not a temporary thing. She describes her bad mood as a tsunami-like depression that she is forever outrunning. If she doesn't outpace her depression with mania, she tells me, she fears she will be engulfed by its dark swollenness and die.

I work with a whole new generation of students who share a trait of affectlessness, or at least for whom moods, good or bad, fail to register on their faces. Are their studied deadpans simply a manifestation of screen face or computer pall? Do they have interior states they're wary of revealing, or is that state in techno-experimental transition? They brighten to the presence of my twenty-first-century mood ring, noting and commenting on its rapid changes more than on the import of what I might be saying or have said. In this way, I know, they do have feelings and can rise to the occasion of a mood. But mood rings are deceivers as I've hoped to show: my current 99¢ job gets more compliments than any real jewelry I might own: "Where *did* you get that ring, dahling?" "I've never seen such a beautiful ring, Miss!" "Is that ring a family heirloom?" are some of the comments the bauble draws forth. I can pretend not to believe in its wiles, but like most people I maintain privately superstitious quirks that are best left unshared: for years I wore a triangular-shaped ring with which I sometimes performed a private ritual: treating the ring like a compass, if it were "balance" I felt I needed on a particular day, I wore the ring with its uppermost point turned sideways; if it were a good mood I was in search of, I'd turn the ring to point up.

If we detach ourselves from, listen to, and pursue our subject in equal measure, maybe a truth we could not have imagined will come to light: only by essaying mood rings and inclinations could I conceive of an option obscured by mourning, blunted by unbearable pain—the memory of my young aunt Frances coughing blood into a respirator tube, of dialysis draining but not replenishing her, until she died; only now can a consideration offer itself to view: that her kidney disease was the result of the administration of lithium, whose prescription was tantamount to an admission of a mental stigma she could not bring herself to share. Why, elsewise, would my uncle have refused to pursue an autopsy to help explain her untimely death?

No more probing; no more application of salves.

He knew that she knew that we could never know. He granted her the freedom to take the mystery of her unhappiness with her to her grave.

they're playing our song

A lot has been written about music and mind but what about music *as* mind. What type of Rorschach would it be to free associate a type of music to each person in your life, or even to strangers on a train or in a waiting room? Jean: Thelonius Monk. Mom: an aria from *Aida*. Dad: the racket of a neighbor boy trying to play a trumpet. My friend who tends a wild garden: cacophonous bluegrass. You there: the sound a bowling alley makes. And you: the needling harmonics of a fingernail across a chalkboard. Each of us is a music minus a symphony reliant on strut. Who am I to say you are essentially a garage band? Some people have a more versatile instrument supply: a xylophone in one corner of the mind, a reed organ in the other available for instigating a song in the heart, irreducible to a tune in the head. But wherefore the access to a fluent glissando on a rolltop piano for some while others live the mood of their days as one-note Johnnies? Perhaps we are all tenors without vehicles. You wake and strap on your mood accordion but you are without a town fair or smoky café where vodka flows and kippered snacks abound. The accordion gets heavy after a while of there being no place for it until one day you trade it in for a pair of dented trashcan lids.

A world in which everyone is compelled to try his hand at the zither could not be a happy place.

A Rorschach is not the same as a work of art.

Take the artful cartoons a young artist, only twelve years old, made from her place at a table with adults (fig. 9). Translating each

person into an animal that she sketched, she seemed to catch something of the essence of each. And she was witty without being mean.
She found the right girth and grin for each of our personalities; to
each, she granted the playful patter of a mood, and almost everyone
was lent one or more tiny accoutrements that perfectly adhered to
them like the distillation of a tell.

Jean was a lynx with curling whiskers, one pierced ear, and a tiny
crown. The child had discovered in us a beanie-coptered beagle,
a wine glass–wielding elephant, a bow-bedecked crocodile, and
a wooly yew. Even the waiter—a presumed stranger to us all—
suddenly grew into the decked-out aura of his own singularity—a
maraschino cherry hung from one of his spangled horns (no doubt a
reference to the condescending fruit he adorned the child's Shirley
Temple with) and his animal, unlike everyone else but me, was also
granted a word. That one word was "Gee." If everyone else was their
animal noun—a flamingo, a hippo—I was asked to mesh even more
fully with my animal as adjective: "squirrely." I was also supplied a
great many dedicated details. It was a real study. In homage to my
squirrely-ness, I wore shoes that bore the imbricated patterns of an
acorn that rhymed with a matching choker at my neck. In my right
hand, a book with the defining monosyllabic title: "Poe." Above my
head, a thought bubble filled in with a mysterious content: "Jimmy
cracked corn and I don't care." Whether the child had captured a
mood competitive or high falutin', dumbfounded or laissez-faire, I
cannot say; only that she had gotten something of me right that I
certainly could never be made capable of seeing for myself.

Which may explain, why, in musical-Rorschach mode, I neglected
to ascribe a type of music to myself, failing to admit the music that
comes to mind were I to try to put myself before myself is Gregorian
chant. It's impossible to hear the sound of our own voice; just as we
can't ever perceive ourselves being perceived, we can't really know
the shape of the mood in which we regularly reside. We can't hear
the hum of what encases us, as such, singly, but in yearning we have
occasion to identify a piece of music-in-common as "our song." Different from the "beautiful music" lovers are said "to make together,"

(9. to each, she granted the playful patter of a mood)

"our song" is a sort of defining moment mood-throb-of-a-tune only available through an anonymous "they" who are without desire and whose only purpose is doggedly to give our song to us.

Not all waltzes, however, are meant to be danced to, and this is what I wish to understand. Take Brahms's Waltzes, Opus 39, where two elements within a phrase are uncoupled without, in the process, erasing the terms of a relation. Here are waltzes that don't require or even invite a movement that would make them "our songs." In the small concert hall where I hear them performed with the—some would say intimate, others would say paltry—company of ten other audience members, lights swell rather than click on, and we are asked to clear our slates of everything but a pianist's red tie against the gleaming field of a black piano. The endurance of the virtuoso inspires a mood of laborless labor, while the music itself spins and lulls, chimes and breaks into anything but a ballroom, more like a woods or dark corridor down which harmony chases tumult, and sleep sleeps itself awake. In the park en route, a stranger's voice spoke of pulling back the hammer and the spring. Gun talk. Inside, a different spring and hammer struck. Inside, the program described the solitary player as a product of near-tragic rise and fall, settled now into this mood of consummate reclamation and return. Is this his song, Brahms's song, or the song of an unrecordable moment of this accomplished finger finding the script of that defining note?

When we speak of the tenor of the times, we don't mean Caruso or Pavarotti, we don't anticipate a fellow human who most beautifully intones; we mean to suggest the pervading intonation. I would like, however, for there to be a tenor for these times. The sweet voice of a man pitched high up, higher than the eye can see. When I try to think of who the tenor of these times might be . . .

mood questionnaire

Please record your answers to this questionnaire on a colorful post-card of your own making. Please pronounce your responses into a tape recorder and play back the answers to coincide with birdsong in Central Park. Please write your answers in a micro-script suitable for homemade fortune cookies or, alternatively, on a very long sheaf of papyrus running end to end, preferably overtaking the bills table, creating a new favorite room in your house. Time your answers to coincide with the slow or fast movement of clouds on the day you are composing them. I am happy personally to receive your answers on the condition that you record them on a piece of light yellow paper in blue fountain pen sealed in an envelope with a wax stamp in red. Another option, following that of most writers, would be to fold your replies into a bottle and toss into the sea. No responses via e-mail, please:

☁

What mood are you in right now, and how has reading aided or abetted your atmosphere?

☁

When in your life did you first become aware of having a mood? What conditioned your, shall we say, mood-day? And do you think we should mark this occasion in the way we do our birth-days?

What, if anything, are you always in the mood for?

How good are you at sensing other people's moods? Are your own moods legible to yourself or others?

Would you be likely to purchase a wall paint that would sense your moods and change accordingly? (I either dreamt this or heard about it on NPR.) Do you think that rooms can sense our moods or that only persons and animals can?

How about plants?

Does the mood of a nation have more to do with the country's economy or its weather? How are the two related?

Has life ever made you literally seasick? Is seasickness a mood?

Are you aware of times in your life when your mood radically altered the course of events?

When was the last time you felt part of a collective mood?

Describe a mood you were only ever in once and have never been able to re-create.

❧

Do you believe that temper, temperament, temperamentality, or temperature affects our ability to be creative? How? Do you think that bile or biliousness, in particular, infects the creative juices, fueling or disrupting the creative process? How do you decide?

❧

What is the relationship between being in one's element and being in a mood? Is our element that which suits and defines us? The atmosphere in which we are most fulfilled? Or is it more like a gemstone delivered at birth? Does everyone have an element? Why is it not possible or what would it mean to be perpetually in one's element?

❧

If you could see your life as a structure of feeling, as a spectrum of colors or a musical scale, if you could chart your life's seasons in the way of a farmer's almanac, what would it look like, and how would it sound?

❧

Does the situation of twenty-first-century Being call for a "mood room"—a space that could intrigue us enough to venture into it because of the promise it might extend to embrace, surround, or hold, if not exactly swaddle us, for the prospect it might offer to enter a *state* of feeling, the condition of possibility for a feeling, if not a particular feeling per se? Would a mood room feature candlelight or neon light, the scent of apricots, the color of sky? Would it feature books splayed on lawns or ardent rustlings; amber glasses and argyle socks; clavichords, b-y-o-beatnik bell-bottoms and electric guitars; the swell of crowds, or the sound of spades, the turning of soil or the slapping of cards, the hay in the barn?

An answer to a modern problem, a question mark curling round the shape of a modern conundrum, possibly the product of modernity itself, a mood room can be effected by the beats inside of paragraphs, the advent of an altered state inside a book, of course, but also *without*, in a place where words can know another, separate, off-the-page life, comingling with the strings plucked by philosophers and architects, filmmakers and scientists, psychologists and musicians.

Have you ever visited David Wilson's rare and wonderful Museum of Jurassic Technology in Los Angeles? To me, it is a mood room par excellence. In a sound portrait of this magical place, Lawrence Weschler mentions how people who wander in the rooms invented by David Wilson either laugh uproariously or cry, or, we could add, become absorbed by feelings we have no name for. I remember the rooms with such immediacy at a distance of months or years: someone will be playing ancient instruments, or folk, amid the red-gray smoke and scent of braziers on a rooftop courtyard studded with archways and coves. A greyhound glances by inside a hallway that also sports a Russian tea caddy. You don't look at what's in front of you here but through pince-nez projected through a microscope displaying tiny still-life mosaics, athwart the room filled with nothing but blue-tinted X-rays of flowers, beside the space marked off by miniature windows of trailers in parks lit by full moons pierced by the bark of coyote, rhymes with peyote, but you are not high on 'shrooms—you're awash in atmosphere.

Would a mood room be nothing more than a definitional play space? Come play with me. William James, according to philosopher Richard Shusterman, was in pursuit of something James called a "strenuous" mood: recall that constipation ran (or failed to run) in the family. Our indices of mood determine everything, capable of turning us every which way, just as they reflect, in what they foreground or forget, the preoccupations and holding patterns of the cultural moment in which we live. For James, there are serious, carefree, easygoing moods and reverential, fatalistic, and uncomfortable ones; there are "feminine moods of mind," "emotional moods," and "mental moods," though he does not ever really define them. Using a contemporary made-in-China desk calendar that calls itself *The Daily Mood: Tell the World the State You're In*, American office workers are invited to flip a page to display their current mood from their place in a suite of cubicles, but the very fact of such a thing is all-around depressing. To each boldfaced state, there's a matching emoticon for "addled, listless, puzzled, quixotic, crazed, rockin', redundant, focused, copacetic, contemplative, cantankerous, antisocial, hunky-dory, neglected, misunderstood, and nonessential" moods. I could add apple butter, spiteful, spice-full, ridiculous, zany, puffed up, recalcitrant, itinerant, nose-picking, cozy, nosegay, fruit-fly, fruit-flavored, psychedelic, hyperbolic, supercalifragilistic mood. Mood is either so limited in our lexicon as to blunt consciousness into oblivion, or so infinitely possible as to be meaningless. Mood is alpha and omega, it is everything and nothing, it is SniffNGo. SniffNGo is a "mood management tool" whose motto is "manage your moods, whenever, wherever." The company offers four gel-stick scents that you can carry in your pocket or in your purse to match the mood you have a hankering for in a range of parts of speech: "sensual, awakened, energize, focus."

If mood defies representation, we should be protective of it, for it must have something to tell us that we cannot see and need to know. If mood defies representation, we should be suspicious of it because that must mean ideology is working in its name. For the pragmatists who founded an American philosophical tradition of that name, moods maintain the sense of being both beyond our ability to know them and "the deepest determining principle" of behavior. In a pragmatist aesthetic light, moods are both the states created by works of art and the nebulous origin whence art comes. In a more daily way, moods attract like-minded feelings to their sphere, blocking our access to others until they take on, for William James again, that great theorizer of emotion, America's first teacher of psychology, the cast of something we need to overcome. Reading a pragmatist account of moods, I begin to imagine them as self-generating parasites that need to be eradicated, each person's quicksand or tarn from which hope cannot spring eternal, with the pragmatist the man in search of a method for mastering his moods. Can you master a mood with-

out setting something on its surface—a pebble, a feather, a word—
and waiting for what sounds? William James seemed to think you
could behave a mood away by purposely acting in ways it doesn't
want you to.

Not entirely willing to disagree, I beg to differ: to be open to
mood is to be open to influence. I might have to give in to this mood.

in a studious mood

Once there was a poet who loved it in his room so much that he spent the whole day sitting in his easy chair pondering the walls before his eyes. He took the pictures off said walls so as not to have any diverting object disturb him and lead him astray into observing anything other than the small, nasty, grimy wall. —Robert Walser, *Six Little Stories*

Thereafter, day after sunny day, through July to late September, [Alexander Graham Bell] smoked new plates for his apparatus, shouted and sang into the membrane of the human ear, and dashed downstairs from time to time with a new tracing to show. His mother watched him anxiously. . . . But there he sat in his hot little bedroom under the eaves, shouting, e, ah, a, by the hour, and, though he did not disturb her, because she could not hear him, she wished that he could be persuaded to spend more time out-of-doors.
—Avital Ronell, *The Telephone Book*

Power makes men mad, and those who govern are blind; only those who keep their distance from power, who are in no way implicated in tyranny, shut up in their Cartesian poêle, their room, their meditations, only they can discover the truth. —Michel Foucault, *Power/Knowledge*

It doesn't begin: "Now is the winter of our discontent made glorious summer by this sun of York."

Nor does it begin: "During the whole of a dull, dark and soundless day in the autumn of the year, when the clouds hung oppressively low in the heavens."

Nor, thusly: "There are some strange summer mornings in the country, when he who is but a sojourner from the city shall early walk forth into the fields, and be wonder-smitten with the trance-like aspect of the green and golden world."

It doesn't open: "Midway on our life's journey, I found myself in dark woods, the right road lost."

It doesn't start: "It was a summer's night and they were talking, in the big room with the windows open to the garden, about the cesspool."

It's neither a play nor a poem. It isn't a novel, though it does create its own architecture of words, brick by brick by brick, tow-eringly so, and dense. No wind, rain, sun, or snow seems to exist in its world; it presents itself as though it were without milieu—or moodless—and yet it creates a mood in me.

It begins: "The ear includes three anatomical dimensions, namely, the outer, middle and inner ear. The latter consists of the auditory cochlea, the vestibular utricle and saccule, and the semicir-cular canals." It's a chapter in a textbook edited by a founding prac-titioner of otolaryngology of whom I'm fond, but that's not why I insist on reading it even when it resists me at every turn.

I like writing that resists its reader; I'm suspicious of the easy invitation that bows to protocol, or the stuff that chatters recogniz-ably, incapable of interestingly interrupting my day by making my heart skip a beat or requiring that I listen with my eyes.

This chapter doesn't merely fail to invite me into its precincts; purely and simply, it neither needs me nor wants me, which is maybe what makes me wish to *study* it—that, and the fact that I maintain a fascination, starting with the three little bones (or auditory os-sicles) that lodge in the middle ear, shaped like tiny tools, as useful as they are beautiful, crafted by an ancient carver and later dug up at the archeological site that is this human body, to say nothing of the chambered nautiluses we all carry around inside our heads, and the thought of how delicate an instrument the whole auditory system is, to this day, a total view of how exactly or precisely hearing "happens" still beyond our understanding. We know it's a matter of translation of something entering in one form and being conveyed into another after passing through two openings on the sides of our heads, not necessarily simultaneously, sometimes with one overpowering the other, and sometimes just through the bones of the skull; we know that filaments grouped in bundles in perfectly ascending rows wave like tentacles, receptors, vital and fine, on the bottom of an ocean floor—it's almost impossible not to metaphorize what goes on in there—and that a relay happens between the mechanism inside the ear canal and the brain by way of an auditory nerve. We just don't really know exactly how the various translations and distinctions are made that manifest as interpretation and perception.

We know that the ear sends sounds out even as it receives sounds coming in. That our outer ears are not likely to move to follow sounds the way an animal's ears do; nor do they sweat; that earwax is a type of inner sweat: the glandular secretion of modified sweat glands. We know, or try to. We write, we read, we listen, we know more or less. We say and say and say. We point. We cut. We chop. We examine. We read in a studious mood.

Because how I hear has a lot to do with how I feel; because if only I could train myself to listen better, I could have the experience of many people, many moods; because sound acts upon my mood by way of hearing, and hearing modulates my mood; because I find the idea of listening being dependent on a physiological labyrinth magnificent and strange, I agree to read a writing that is unreadable. Having evacuated the need for an addressee, this writing is its own object: a mise en abyme of naming of parts, it sets in motion the possibility of segmenting the ear, and with it, audition, ad infinitum. But it's not this hysterical aspect that makes me giddy while I read—giddiness being an unlikely effect whilst seated before a textbook and yet here I am experiencing a perverse excitement with each uniquely opaque turn of phrase. It's the way in which the language assumes the possibility of something better and more than jargon. It's the way in which I come to understand the chapter as a replacement ear, the writer having greeted the body by turning away from it or by supplying a body made of words in its stead; the doctor-anatomist facing the challenge of the ear's unknowability by creating a language, bony and protuberant.

Certain sentences must be read aloud to glean their full effect, as in: "Anteriorly the cave of the concha is bounded by a tongue-like projection, the *tragus*, which is separated from the antitragus by a deep notch, the *intertragic incisures*." Or: "The concha is partly divided into an upper portion, the *cymba conchae*, and a lower portion, the *cavum conchae*, by an oblique ridge, the *crus of the helix*, which is continuous anteriorly with the margin or *helix* of the auricle." Or: "The external surface of the tympanic membrane receives its blood supply from tympanic branches of the deep auricular branch of the internal maxillary artery, while the internal mucous surface is supplied by branches from the anterior tympanic branch of the internal maxillary and the stylomastoid branch of the posterior auricular." Or: "The chain of ossicles, with their articulations, forms a series of levers by which the movements of the tympanic membrane are transmitted through the base of the stapes to the perilymph of the labyrinth." Or: "Clefts between the bodies of these pillars enable endolymph of Nuel's space to communicate with that of the tunnel of Corti."

Enter this foreign land rendered in a foreign tongue to describe something as intimate as the ears that are plastered on, impossible to dissever from our heads, and discover this writing's tantalizing tensions. Here we have language wielded with X-Acto knife precision whose effect is a prose that is thick as mud, and obdurate.

Though apparently meant to yield and give way, to impart a kind of knowledge to us, to deliver factoids unto us, sentences repeat and fold back on themselves rather than propel forward. It is partly this aspect that might make us feel that we are in the presence of the most disarming, because unself-conscious, nonsense.

Like poetry, the writing works more at the level of the "signifier" (or sound) than the "signified" (the meaning to which sound points), but the difference is that, unlike poetry, this language pretends to be entirely referential, even transparently so. What it names, it thinks of as real, and the lengths it goes to name and name and name (and therefore, claim) belies the body's squirm, recalcitrant and irreal. Insofar as it's a writing of a body that seems to absent the body, we can't help but be startled by the sudden mention of the living human form at that point when the author suddenly remarks how the lack of fat cells in the outer ear makes it "especially susceptible to frost bite"—then, the body-as-sensate returns as if by mistake, calling for a warm set of earmuffs or a hat.

Like all anatomies, this linguistically dogged flight of fancy has in mind (without having to say so) dissection of the dead body as precursor to surgery on the living one, but its determination to segment and isolate, to divide and conquer proves difficult if not futile: the impossibility of sequestering its subject so as to study it is proved by the writers' persistent reliance—beautiful and strange—on that most situated and relational of modes, on metaphor.

Parts of the ear are "jelly-like," "ribbon-like," "bell-like," "tunnel-like," "ivory-like," and "calyx-like." Landmarks are named for their discoverers or conquerors; encountering them is a bit like finding men's names marking the beyond-our-view, extraterrestrial territories of outer space: there's "the organ of Corti"; "Deiter's cells"; "the space of Nuel"; "the cells of Claudius"; "Prussack's pouch"; "Shrapnell's membrane"; the "tubal tonsil of Gerlach"; and the "stripe of Hensen." The inner ear is likened to a cityscape—it has "aqueducts," "canals," and "tunnels." One minute it's an aqueous ecosystem supporting an "isthmus," and the next it's a pastoral landscape with "twigs" and "supporting hillocks." I favor the language that likens the ear's complex physiology, its structures and shapes, to a house: there's an "attic" and "roof," a "sinking floor," and "wall," there's an "oval window" and a "round window." Lest we think we're in a humble abode, there are "pillars." These images—magical in themselves for wondering a house with windows inside the dark space of our heads, also make it possible for adult, mostly male, doctors to play inside a dollhouse they were in infancy denied.

~~~~~~~~

When a philosopher I like to read—Giorgio Agamben—in an essay as spare as it is voluminous on "The Idea of Study" draws a line between study, stupefaction and stupidity, he must have me in mind, for no matter how turgid the prose—and perhaps the more turgid the better—when I'm in a studious mood, I agree to chain myself for as long as it takes, the better to endure whatever that which bores has in store for me and thus my willingness to list and drift and hold steady for entire afternoons inside my anatomy textbook's ornately brocaded rowboat. One doesn't, as one might suppose, study in order to be made less stupid, but to stew in stupidity's juices for as long as time and life allows. To defy the time allotted to one. "Those who study," Agamben writes, "are in the situation of people who have received a shock and are stupefied by what has struck them, unable to grasp it and at the same time powerless to leave hold. The scholar, that is, is always 'stupid.'" He goes on: "This shuttling between bewilderment and lucidity, discovery and loss . . . is the rhythm of study." Temporally speaking, study is interminable; it knows no end.

The studious mood sets in motion the perseverance of the little engine that could. Sometimes I think the studious mood will put up with anything. Its initials—S/M—set the terms of its contractual agreement, otherwise how to explain my need to finish reading sentences that are in every way ungraspable, that yield so little but hold me fast? In studying, do I succeed in creating a barrier around myself, much in the way I learned successfully to concentrate on homework and thus to craft a room or hermetic hollow I could call my own inside the ever-noisy, never private mayhem of the row home I grew up in? My bedroom was too small to hold both desk and bed, so I carried out my studies on the dining room table amid TV blaring, father yelling, mother singing, and brother practicing a piano. Studying could be its own white noise—it blocks out one set of stimuli so I can hear another, far off, unheard tone meant just for me so long as I give to it my supreme and attentive patience.

To study is to be chained to a book; to read is to be released into and through a book. Studying, we hunch over a book. Reading, we take it into our lap like a favorite pet and curl up with it. To study is to peer; to read is to gaze. To study is to focus; to read is to graze. I wish the distinction were this neat and easy to assume, but it's not.

Etymologically, "study" ranges from a state of reverie and abstraction, to an attitude of friendliness or devotion to another's welfare; from zeal and desire to an inclination or a pleasure felt. From a preliminary sketch to a representation meant to reveal an essence born of careful observation. From a cupboard for holding books to the room that holds a person reading them.

I begin to wonder whether study requires an object or if it is essentially a place or a state. There's that type of study I yearn for in a phrase my grandfather would use to describe an early morning activity: "to inspect the garden." I know what it is to follow the call that combines noticing with roaming, tending with attending, passive pleasure and surprise with lordly purveying of what one has made, most of all at a pace that is exceedingly *slow*, in which time is replaced by shapes of tendrils and hollows, of leaf and furl and breeze.

Then, too, there's the greater lesson buried in the type of study that is itself a lesson—the beautiful restlessness of musical études where what begin as mere exercises, music composed with the express aim of conveying, via practice, technique, morph into some of the most complex and dreamy scores ever written.

~~~~~~~~~

In what sense is "study" both as noun/place and verb/act constitutive of a mood of the past, a zone of the present, or a future atmosphere? At Boston's new performance space, Calderwood Hall in the Isabella Stewart Gardner Museum, I learn of a new type of music-listening opportunity called In-and-Out Concerts. One can't blame beleaguered cultural institutions for trying to find ways to entice a busy public in, but the name of the series bears the unfortunate onus of equating music-listening with unsatisfying sex, fast food consumption (it's the name of a burger chain in California), and one-stop shopping. "Experience rarely performed pieces in a leisurely come-and-go atmosphere," the ad encourages in a kind of oxymoronic formulation as guarantor of bourgeois freedom: if leisure suggests you have time apart from work, "come-and-go" reinforces an earned complacency, the opposite of needing to "show up" or punch a time clock, but it also implies one's being chained to a treadmill whose speed one can't control since it's understood you might not have time, or the interest requiring that you make time, for a full concert. Such concerts don't discriminate between audiences, the ad goes on: "Enjoy a few moments of musical inspiration or enrich yourself with a complete performance."

Imagine giving a reading from a book you've written to an audience that comes and goes, catching a snippet here, a word there. I think I'd be tempted to wave hello and good-bye to each passerby; to throw the book out with the bathwater and revert to a type of performance art focused on greetings and good-byes. I'd ask someone to put a button in my back, which, pressed, would make me more like a marionette. I'd close and open my eyes more slowly than any human might, and for each person entering and exiting, I'd supply a different sort of wave. I'd make of myself a study. I'd do this for hours on end until something would reveal itself in the chink between my index finger and thumb: those only capable of appearing for a nano-second would experience the goose egg of inspiration; those who stayed would be enriched by the shape of an emptiness they hadn't known they were in search of: in a matter of minutes, they'd fall fast asleep.

But let's get back to anatomy textbooks and the student, like myself, who "sits in a low-ceilinged room 'in all things like a tomb.'" I sometimes think my studious mood bears the qualities of a beneficent faith, convinced as I must be that if I lend enough devotion to my task—not exactly obediently or even routinely but in the manner of a naive follower—something will emerge that I could never have anticipated finding. In "Anatomy of the Ear," it comes in the form of two sentences very near to the end of the text: "The antero-inferior quadrant is strongly illuminated when the membrane is examined by reflected light, and is known as the 'cone of light.' From the apex of the cone, at the umbo, the shadow formed by the handle of the malleus may be followed upward to the malleolar prominence." Needless to say, there is no cone of light inside my ear unless a fantasy writer passing as a scientist puts one there, but the idea of it repays my studious mood enough to keep me studying. This is not an example of what is referred to in contemporary parlance as a "take away." It's an unintended bestowal of a poetically imbued arch. It's a sign of the conjoined frailty that brought the anatomist and myself to the same table. It's proof of a shared pathos that needs no justification. "Umbo" is such a beautiful word, reminiscent, to my ear, of Italian, "ombra" or shadow, but nearer in meaning to a rounded knob or protuberance. You shine a light into a cave; you give it a name based on the world you think you know; you follow the shadows thereby cast; you're in search of the origins of a music tuned to the weather in a person's head.

Maybe a studious mood is a call to darkness—a willingness to climb inside a cave—and the accompanying challenge of how you will make light happen there. Midnight oil; candlelight; moonlight. We tend to associate the mood of study with fading light or light that requires stoking or the slow and careful tending of a wick rather than with the stark unmasking of break of day.

My own most recent study on the top floor of our house in what once was an attic had been beginning to feel too dark to me. When my father died, and I received the sum that was equivalent to my third of what his row home was worth—ten thousand dollars—I didn't have to think for a minute about how I would use the extra cash. I decided I wanted for my father to bequeath me light; I could receive from my father something he may not have known how to give me when he was alive, but that I could still associate with him. Not wishing to squander the money he earned working dawn to dusk shifts for forty-five years as a sheet metal worker at the Philadelphia Naval Shipyard, I sought a way for his bequeathing to have lasting significance. I decided to have a skylight cut into the roof right above my desk. On the day the contractor installed it, I felt compelled to tell the man who carried out the job how work begot work; how this wasn't just an indulgence that every day could afford—just another skylight—but I refrained.

I don't know what I thought would come through this window.

The hand of God? A glint of passing cloud? A suffusion of rays that would obliterate the need for lamps and transform my study into a meadow? But since it's been installed, I almost never look up from my desk to gaze through it—the studious mood, after all, is more reliant either on an eye-level window or on staring into space. What's worse, in my attempt to introduce a vault of heaven into the scene of my toil, I hadn't considered the way the new skylight would alter my study's climate, making it hotter in the summer and colder in the winter.

Just as the studious mood is pitched to unexpected inklings rather than predetermined ends, so the construction of what we *think* we need in the creation of a space in which to study is often better built from accident than from self-perceiving need. I've always considered a door that closes to be paramount to a study; otherwise, the presence and call of other people is too readily available to break or disable my studious mood.

Recently, a rearrangement of our three upstairs rooms has led to my dwelling in a more open plan. In addition to the weather that more readily enters to bake or freeze me depending on the month, my chamber is at the top of a doorless staircase that abuts a bedroom, a guest room, and a bathroom. Come to think of it, it's a kind of thoroughfare, especially for sound. In a studious mood, I think I need to be closed in, but in my unbounded room, I hear Jean's presence differently as, for example, today when her arrival is conveyed by three sounds, real and pure and true: her footsteps coming from wherever she has been as light or heavy as a mood ascending a staircase, as much a rhythmic signature as a voice; the "puh" sound of a wooden door stuck to its frame by humidity pushed open with less effort than usual—she must be tired; the dropping of a set of keys onto the dining room table, but they could be marbles, dominoes, or polka dots in the roundly scattered spray of sounds they make.

The sounds are of Jean being-in-time, with or without me; they aren't the sounds of the arrival of the beloved to a lover tortured with waiting. They're parcels of a self, not asking to be overheard, of Jean's separate privacy of being. Maybe from here she will enter her own study rather than greet me—or me, her—but the three sounds strike me as more beautiful a gift than anything I've studied today, though a beauty made available by listening in a studious mood.

They remind me of a day when I was helping Jean look for a lost passport in her study, of looking into a drawer and being struck with the feeling of finding someone I did not know there. I remember being surprised by the nature of the things she saves, and the particular geometry of the way she layers them in her desk drawer, and a sense of gladness to know I could feel their texture so organized NOW, and know her there, rather than one or the other of us, beset with the task of emptying each other's drawers at death, discovering THEN the person we thought we knew—a particular version of the other we'd somehow missed or never loved. We shall not know each other entirely; we will never be able to say we loved each other enough. But such discoveries open a space to make a relationship into a venturous garden with an opening here, an untrod path there. And they serve as a demonstrative reminder: that what I experience, see, hear, come to know or learn in a studious mood may have nothing to do with my object of study.

～～～～～～

Today, perched on the edge of another late-in-coming spring, I'm inspecting the small garden I keep behind our urban house. I notice forms coiled as taut as an infant's fingers and as likely to uncurl reflexively with life. The white petals of a pear tree shower us almost as soon as they bloom, jubilant and momentary as confetti. Am I really willing to pull up weeds this early in the season? Having waited so long for hints of new life, what compels me to neaten the garden's forceful plan? There's so much afoot here, with me or without me, and so many questions as to the origins of signs, the need for identifications, I have no way of knowing whether the things I find in a studious mood have bubbled from below or fallen from the sky.

White flecks, for example, interspersed with earth strike me as calcifications, bits of broken pottery, or chipped paint that's drifted in from a neighbor's flaking siding. A perfectly round object half-hidden in earth could be a worm perfectly coiled, I think, then again a part loosed from a hundred-year-old piece of machinery. To tell the truth, I thought it was a cock ring. Plump bumble bees move in pairs athwart the hose spray; the rhododendrons' leaves are drooped and charred, scarred perhaps by the harsh winter, so I feed the plant until I overfeed it, which will no doubt stress it more. There is damage and blight and mystery: the tops of the cardinal flower plants, seen growing in riverbeds that bring bright red to autumn, appear chopped and mowed, but there's no sign of a literal scythe in sight. Squirrels have gnawed the tops of begonia plants I kept alive all year inside and uprooted a geranium I'd overwintered. A plant with heart-shaped leaves that didn't bloom last year reappears, or maybe I hadn't noticed it. Tangled inside the honeysuckle bush, angled against a wooden fence, a long and trailing piece of orange-colored yarn.

At first I ignore it, considering it on a par with plastic bags impaled on trees, a piece of trash that might have floated in, until, bending near to the plant, I'm startled by a fluttering: the small body of a bird darts past, evidently having exited from a birdhouse hidden in the honeysuckle's leaves. Sometime in the 1950s, my father had built the birdhouse—a kind of guest chalet with pointed, scalloped roof fashioned out of scrap sheet metal—in the hope of luring birds to his own garden. In the process of dismantling my father's own house after his death, I asked my brother if I could have the birdhouse. It was ungainly he reminded me, much heavier than it appeared, and crawling with insects: did I really expect him to send it to me through the mail? I did, and he did, and Jean and I mounted it: in fact, it fit perfectly atop a rusted bedpost that served as a moor for my climbing plants.

The thing about the birdhouse was that it had never attracted any birds, even after my father experimented with engineering a smaller opening that he superimposed atop the too large one he had started with—an improvement, I recall, requiring welding and rivets; even after he'd tried to offer the birds an anomalous wooden dowel as perch that he affixed with glue to the metal. My own theory was that metal was too cold a substance to attract a bird to nest in.

Now I found the yarn, the very color of the honeysuckle's future blooms, delightful, and considered that it had been dropped by the bird en route in the process of making a nest, maybe the very first bird to use my father's birdhouse. I retreated behind a window, so as not to disturb the bird, and continued to study it: it moved cautiously among branches, carrying something in its mouth: a white piece of notepaper, a feather, a guitar pick? It watched, it waited, it hopped, then quickly entered the house again.

Was it enough that the thing my father built eventually worked as he'd intended it to, whether or not he was alive to see it? Imagine what will be born from the bird's movements to and from the seclusion of its darkened room, piled with slips it finds, woven with instinct and with industry. The golden thread, my sighting it, began to feel like a blessing, but not in itself alone. The blessing was in my witnessing it—a witnessing not borne of will but from the accidental browsing of a study in the garden.

But what am I saying? It's not a piece of yarn at all. It's a golden thread unloosed when some Rapunzel let down her hair. Its crinkled pattern makes it most recognizable, identifiable, traceable to a source: it's the lost remnant born of all that tugs on a girl uncertain whether inventively to exit or furtively to craft her mood at the top of a stair. It's a piece of a yearning in progress, adrift and detached, with a life of its own.

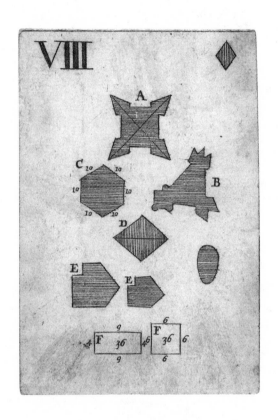

(rooms)

The movie spectator could easily appropriate the silkworm's motto: *inclusum labor illustrat*; it is because I am enclosed that I work and glow with all my desire. — Roland Barthes, "Leaving the Movie Theater"

life breaks in

We all have our synced and un-synced tableaux-bearing memories of moody immanence, perching us, as they so often do, before screens or boxes, rooms or windows, and isn't the reality we navigate built of such translucencies? Overlapping, blocked off, layered, bluffed, contracting and receding, blotted, blurred, clarifying, lit from behind or above, suddenly darkened, but never entirely blank?

At one point this summer, on a cool but not yet chilly evening, I was looking into a set of glass doors from the deck of a little log cabin where Jean and I were hosting a gathering of her extended family in Maine. For that moment, the quiet had reeled in all of the day's competing castings and high energy, and everyone had found their own envelope of attention: we were dispersed and together all at once. Now all of the adults—there were seven of us—happened to be hanging out in the dark with nothing but a candle to light the scene, and from where my chair was set, it was wonderful to see, panorama-like, inside, what suddenly struck me as a "mood of life." Inside, one pair of legs was thrown over the back of the futon, play-style (college-age Justine's), while another set of eyes was trained on the little TV set with built-in video that usually goes unused—now there were images of *Aladdin* showing on it to a mesmerized three-year-old Sam. Seven-year-old Sophie, meanwhile, was seated at a table working to decode the messages we collected on rocks and twigs and bones found during the course of her stay here.

I wasn't seeing this scene on its own—that would have made it meaningless—but by way of other transparencies in a past-present-and-future slide show. I was comparing the atmosphere of our house suddenly populated—all of its members dreaming and playing—with the way it is differently filled when Jean and I inhabit it together, writing in separate corners of its rooms, pausing to prepare a meal or tend the garden; imagining the difference of being here alone with no one but my cat in projected weeks ahead: how I'll start, then, to the sound of a neighbor's dog, or the unusual knock on the door, I'll pitch inside the sounds of the house settling or the scuttling of mice in the walls, and how all of the objects in the room in which I'll write will resolve themselves back into the mute indifference from which they had sprung. I'm feeling, too, the difference of these rooms being someday without any of us, even if they should continue to stand.

The thought lasts but an instant—I'm not allowed to linger there for long—before life breaks in, again.

mood rooms

Indoors or outdoors, we are always *enclosed*: this is what is meant by mood. If moods are inaugural and primal, then we might grant them the condition of a crib or a cradle, devices that keep us from straying or sway us in place. Told to sit still, we set out into the territory of daydreams; rocked to a still point of sleeping, we wander among bedrocks—building for, and from, our brain space the dimensionless rooms we call moods.

I could say that mood is like ether—an invisible gas that makes you laugh or lays you low—but that's only because, when prompted to think of moods as rooms or vice versa, the first thing that comes to mind is an operating room with myself as patient. I peer through a portal, but without the faculty of sight, at four, masked and hatted figures, sure, short-sleeved, and bibbed, who stand at a vanishing point that is actually the near end of my six-foot-long bed. Their main detail is this: they all have their arms folded when, really, they are meant to get down to work. The team of them, green and glowing, seem bored but not impatient, detached but not indifferent, and this combination sets the tone: they are waiting for me with all the time in the world even though the drug they've administered is so quick acting as to put me out in a blink. They are waiting for me in some other world that I need to arrive at but cannot be present to.

Certain rooms from our childhoods are telltale with mood; in my case, not the ever-suggestive basement, but our row home's inner front door. Not a room then, but a go-between. Just the fact of

the door, snow-colored and closed against snow, open to sun showers, pollen, and thunder. Sometimes I stand before the outer screen,
arms folded, musing: the door as egress to unintentionally reflective
moods, an occasion to study a bee or cicada stuck to the mesh. The
space between the doors, a repository for a newspaper rolled like a
blanket inside of which lay the Sunday comics in full color against
black and white, makes me think that seemingly nondescript door
could set or alter the tone of whatever was going on behind it. Otherwise, the idea of mood rooms takes me to closets—small enough
to hold one person lost inside wool coats in a game of hide-and-
seek, or a larger, free-standing one made of cedar that contained my
mother's voluminous wedding dress and my father's diminutive collection of bright-colored polyester ties.

Childhood's most comforting rooms are an autumn-tinged, fold-
out paper village that my mother displayed at Thanksgiving, and a
genre of pop-up books whose paper pull tabs, levers and springs,
sliding windows, and accordion doors perpetually surprised me
even when I knew what they revealed. Outdoors, on a trip to Philadelphia's art museum, I salute the Brothers Peale as they coyly climb
their trompe l'oeil staircase, entryway to a room I cannot see; indoors, I enter books as if they are rooms: I pull on strings and watch
books' rooms unfold in three dimensions.

Such dreamy interiorities are only antechambers to a vast and
variegated architecture of mood rooms that have shaped me, I am
sure, and they make me think I'm mis-concocting mood as a place
we go for refuge. I'm remembering now the perpetual mood of my
father—the poor guy was always in a "lousy" mood, which isn't to
say he appeared as though lice were sucking the life out of him. He
didn't seem depleted or depressed; he just always seemed to be in
a mood that could only be described as "filthy rotten." At the dinner table, he's sour; in his garage workspace, he's sullen; in the bathroom, he's pained; in the living room, he's tortured; before the wheel
of the car, he's anxious. Only inside the menacingly sudsy clutch of
a car wash does his perpetually disgruntled mood state lift, and it's a
curious privilege to be asked along for the ride.

I always agree to join him in this trapped, mechanical venture. We don't talk but sit silently and stare straight ahead as though anticipating the novelty of a drive-in movie but without the snacks. The car may jerk and rumble on its tracks, but my father appears relieved, as though the Jack-and-the-beanstalk-sized sponges were giving *him* a deep cleaning and not just the wheeled chamber that is the car. The car wash's pulsions and hissing hydraulics are evidently soothing, while the long strips of rubber that lap the car could be reminiscent of the beaded curtains that served as vented doorways in that country my father's family never spoke of even though they smelled of its partly bitter, always pungent, braised escarole, fennel, and capers: the island of Sicily.

My father and I are encased in metal and glass that rattles beneath the long tongues' gentle slapping, while in Sicily those always colorful, permeable sheaths that front Sicilian doorways let in air but keep out heat, let in light but keep out weather, let in voices but keep out winds and dust, ask to be parted, and hang like hair, dance on the belly of a threshold on a street where nothing happens for days. On one end of the island, near to where my father's family were from, in Palermo, some of the most unusual subterranean crypts in the world: the Capuchin Catacombs (*Catacombe dei Cappuccini*); on the opposite end of the island, an "ear" born of the landscape's myth-infused rock formations and large enough to walk around inside, listening to one's voice come back to one: the Ear of Dionysius (*Orecchio di Dionisio*).

If you wait to be in the right mood to visit the catacombs, you will never go—as chambers of death, they hardly invite an outing, family or otherwise—but on one of several visits to Sicily, I finally felt able to brave the descent and investigate their corridors. The surprise was that the experience was neither Gothically freakish nor ghoulish and terrifying: the corpses, standing, sitting, or lying down, mummified and dressed in the finery or uniforms suited to eras ranging from the seventeenth through the early twentieth century—there were eight thousand bodies in all—were profoundly humanizing. The place was shot through with a leveling irony: even though the dead

were cannily arranged in groups befitting their rank and station—
all the teachers in one room, all the lawyers in another, and so on—
death here was the only rank and file to be obeyed. If one expected
gruesome spectacle, the most spectacular scene to be found was a
corroding button, or the rotting tassels of an eighteenth-century ep-
aulette: we all decay, as do the threads that hold our clothes together
if given enough time.

I had expected to be frightened in the catacombs. I had expected
wanting to flee. Only fictions about death are frightening, presuming
as they do that death and horror go hand in hand. Here, death was
blithely a face-to-face affair, and the atmosphere of the catacombs, a
surprisingly warm and homey one. I hadn't expected to be instructed
in the catacombs, but I was left with an image that I never forgot—
not one that clung to decaying flesh but that had everything to do
with architecture. To each person, one of a few guiding placards explained, a niche might be constructed suited to his or her shape and
size. During one period of fastidious religiosity, in the name of hu-
miliating penitence, living residents of the city would occasionally
visit in order to inhabit their niches prematurely. A bit worse than
being sent to the proverbial doghouse, the act of standing for awhile
in your niche was meant to remind you that you art dust and unto
dust you shalt return.

The catacombs give the colloquialism "to find one's niche" an
entirely new meaning, as if to suggest that the way we fit into the
world and thus feel suited to it, defined, and purposeful, is synony-
mous with the measure of our days, our coffins (plate 1).

In an interview with Claire Parnet, French philosopher Gilles
Deleuze summarizes his proposition that there is no territory without a vector of leaving the territory, that there is no leaving the territory, no deterritorialization, without a vector of reterritorialization
elsewhere. He shares his fascination for animals, like ticks, whose
territory is extraordinarily limited. But then he quips: "Every animal has a world. It's curious because many people do not have a
world." In a final, arresting example, he recalls watching a stray cat
he'd agreed to take into his house when his child asked for it. The

cat had become ill, and as it neared death, it sought out a corner in which to die. There is a territory even for death, in other words, and the animal puts itself into a territory in order, we might say, safely, peacefully, definitively, markedly, properly, to die.

Is it possible for an animal to have a sense of propriety? Is it likely for a person to maintain a sense of the animal?

Our world is parsed by territories, psychic and bodily, political and aesthetic, imaginary and real, circumscribed and vast. Moods are niches that I want to imagine as separable from the containers that take us to our deaths. I want to believe that some niches exist for us to die in, others to live in, and that those (invisible containers) we choose to live in are our moods.

On one end of the ancestral island of Sicily, subterranean niches; on the other end, an open air, larger-than-life, smoothly cavernous ear. Both were starting points and end points in a familial mythology, and both were laced for me with fear. It's only now that I realize their impression as perfect orienting points in my thinking about mood, for I know that what compels me to follow mood and to attend to it, thoughtfully, lovingly, has been some conviction about the power of enclosures, familiar and otherworldly, just as I've always understood mood to be intimately yoked to listening, to the heard world, sound atmospheres, acoustic pleasures or dins, and in particular, to the voices that carry us.

My father's father, the first generation Sicilian immigrant, on his deathbed reported a dream that he was taking a bus to Syracuse. Since there were no family members living in upstate New York—the original members having all settled in South Philadelphia—my grandfather's children were hard-pressed to understand what the dream could mean. An old-timer, upon being presented with the riddle, explained that my grandfather's parting dream didn't have the country in which he'd spent the past sixty years in mind. It was evident that he was journeying to the other Syracuse, to Siracusa, Sicily, the western port dotted with amphitheaters where the plays of Aeschylus, were, in ancient days, performed, the city with the monumental cavern shaped like an ear.

On my own first trip to Sicily in the aftermath of my grandfather's death, I made a conscious decision not to go to Syracuse. If Siracusa was the corner in which my grandfather chose to die, it wasn't
a place I should immediately visit if at all; and when I was ready to
discover it—which I did on my fourth visit to Sicily over the course
of eight years—I might need to treat it less as a trip and more as a
pilgrimage. It happened to be the year of a cancer diagnosis, but my
partner Jean and I had planned the trip to the western part of the island before this news. I didn't know if I'd be well enough to go, but I
was, and like a lot of people, with or without ritual motives in mind,
when we found ourselves standing inside the echoic ear, we didn't
know what to say, what sounds to produce to feel its full effect. Do
you bellow out "hello!" into the void, or amidst its cool, damp trick-
ling, emit an "ohm"? Do you feel more oxygenated or less inside its
pressing heights? Do you fall back on the idea of all sound as re-
cordable, never as ephemeral, by blurting mechanically, "testing 1,
2, 3"? Do you choose to scream or sigh, whisper or sing; do you form
words, or only make sounds to feel the way they ricochet off your
body, creating their own language somewhere between the ramparts
of heaven and the parched Sicilian earth? Our choices must depend
on our mood: Jean, as I recall, sang a verse her deceased mother
would sing to her, then said it had nothing to do with her mother,
only that she liked the song's harmonies. As the words echoed back,
she created an earthenware round: "White coral bells upon a slender stalk / Lilies of the valley deck my garden walk / Oh, don't you
wish / that you might hear them ring? / That will happen only when
the fairies sing."

Do we only say we're in a mood when the mood that is our normal temperature shifts? What makes our "sad," "bad," "good," or
"strange" moods worth remarking, and what name could we give to
the state we are in the rest of the time if not our "mood"?

If we are never without mood, what leads us to identify some
places as mood spaces and others as scarcely productive of mood at
all? In a biopic about Mozart, the narrator explains how, on a trip to

London in which the great composer's father took ill, Mozart, who was only seven, "occupied himself" by writing his first symphony. Our moods are our ways of occupying ourselves; they are our literal occupations (see niches again). When Mozart occupies himself by writing a symphony, he fashions for himself a place in which to reside; he creates a room within a room, a container within a container, a language for a mood. If moods are rooms, feelings are the objects in those rooms; art, their rearrangement.

If you could only name five types of rooms as constitutive of the mood repertoire that creates you as a feeling subject in the world, what would they be? I would have to name fountains and birdbaths, especially in gardens in working-class neighborhoods, even if they aren't exactly "rooms"; kitchens, in movies and in real life; 1960s television, the TV itself a room (or box), most especially, Dick van Dyke's living room; meeting rooms and cubicles; Catholic confessionals; library rooms—especially listening to vinyl in an undergraduate library; reading books as a form of listening, reading as a steam bath or sauna of the mind; leaving a library in an altered state and nearly getting hit by a car; classrooms as rooms a group of people make; and most especially, vestibules—the room that is in the wing, the antecedent to public ritual, visibility, and booming oratory or group-think: the backstage, hidden room of fore-scents and touches of cool water. The town I grew up in—Darby, PA—itself a vestibule to the city I could not see but only hear to the tune of my imagining in the distance.

That's apparently more than five, perhaps a baker's dozen, but who's counting? Mood rooms are many, multiple, and multifarious. Only when we're depressed do all the rooms compress to one—see the sort of sentence that recurs in the short notes I receive from my brother: "The problem is I am somewhat depressed and find it hard to escape from my microcosm."

Some mood rooms recede, while others compete for full occupancy. Some are vast as grass; others as small as a corner, and as isolate. Still, they share this trait in common: all of them are equivalent to how your people held you. Whether you feel trapped in rooms or able in them, whether memory opens its doors or keeps them shut has something to do with that, too. There's that man, my father, not exactly happy, but at least calm inside the car wash's lash and lapping. I'm in the passenger seat, but it's as if we're both driving, and I feel a different sense of him for a spell, being carried along.

miniature verandas and voluminous velvet forms

What knowledge is this that recognizes without ever having seen and knows without understanding? —Gabriele Schwab, "Words and Moods: The Transference of Literary Knowledge"   ·

It comes in the form of a pince-nez or black box with faux telescopic eyelets that I hold before my face. I'm an adult, but this thing is a toy. The square outline of the box is broken by a dip that could be a rest for the bridge of my nose, until I realize, turning it topside, that it's the place to insert a paper disc perforated by photographic images. Holding the device to the light as instructed, I don't see anything in the outside world more clearly or more nearly. I'm invited instead into the theater of a manufactured view. I don't look *at* anything, but *through* in three dimensions.

The images I look through are motionless but suspended, which makes them seem as though they move. And I, too: I'm riveted, and, yet, I bob. The spatial orientation is just as private as a peephole's but more capacious: the boundless circuitry of the child imagination ousts the vapid focal point of adulthood porn. It's the difference between thrall, which knows of atmospheres, and thrill, which needs an aim and end. It's the difference between enchantment and excitement.

I would never say the "scene" (for lack of a better word) or "objects" (for lack of a better word) that make up the scene seem real but aren't real; nor would I say that the viewing device and its view "brings things to life." The charm of the thing isn't one of animation but of captivation. The apertured black box masks things off—everything that's not inside it—and it sits on my face like a mask to my features. The contract is this: I can see without being seen and see only what it wants me to. If this sounds too menacing, consider what I'm being asked to view: hand-painted clay figures wearing elaborate handmade gowns. How can such miniatures bear such subtly expressive faces? They reside in velvet-curtained backdrops; they walk on widely planked floorboards and eat at hand-carved tables. There is sky and cloud through their windows, and a light that indicates a source beyond the alcove of a Gothic castle. Often the figurines gaze inside of mirrors that then reflect their faces back to me as well as unseen details of their rooms. There's a cat and dog in every scene, witnessing and reacting, inside a handmade forest. Among my favorite scenes are those that promise deeper and deeper unseen space, with or without the characters: Snow White and the Seven Dwarfs.

I'm not sure if I ever owned a View-Master as a child—but how can that be when over one billion were sold from the middle of the century onwards and given how especially popular the devices were in the '50s, '60s, and '70s? I seem unable to know if the interest I fasten onto View-Master fairy tale reels as an adult is buoyed by a particular memory of having once been held felicitously captive by a View-Master as a child—in which case, what tethers me is nostalgia—or if, to my adult gaze, in my adult grasp, their hold on me is patterned by some more free-floating flecks of childhood wallpaper unstuck from the walls of any actual occurrence and now suddenly available to my view.

Do View-Masters avail me of a particular species of remembrance—a *mood memory*, let us say, to which nostalgia might serve as impediment, for here, there's no conviction of a return. The View-Master doesn't usher in a long-forgotten mise-en-scène as if opening the curtain on a sepia-tinted family photograph. It's more like, inside a no-space that is a yes-space of untouchable light, the apparatus and its scenes invite the hue of some broken-off part of a mood; it's as though some however amorphous echoes of an underwater muffle that was a (childhood) mood sounds, fades, and sounds again; appears, disappears, and reappears. An atmosphere once held me—I maybe even swam inside its solemnness or swirl—and now it was there again to view, but I had never seen it. Blocking off the world to enter a View-Master's in-scape, I go somewhere I have a hunch I know, and someplace I've never really been.

~~~~~~~~~

I don't exactly "collect" the black Bakelite binoculars, but I keep, in a zone verging on fetish, on a shelf with a handful of perpetually changing favorite books, three 1950s-variety scopes, two defunct View-Master projectors, and approximately fifty View-Master discs. A local antiquarian once called me, intent on selling me a thousand discs that had come into her possession, all carefully indexed by the places they documented (they were the travelogue variety), but I had to explain these weren't the types that fanned the flames of my strange enthusiasm.

Introduced in 1939 at the New York World's Fair, the Gruber-Sawyer stereoscopic device originally aimed to offer an alternative to the scenic picture-postcard, establishing itself by way of monumental vistas—the Grand Canyon—and deep space caves—the Carlsbad Caverns. Two of the company's popular mottoes—"what in the world would you like to see?" and "the world at your fingertips"—give a sense of the View-Master's sweep, for, so long as we're in the land of pictures augmented by 3-D, there's nothing *not* worth viewing: View-Master's subjects could afford to be encyclopedic, so that, alongside its ever-popular snapshots of places near-off and far sits a twenty-five-volume atlas of human anatomy, flower and fauna and beasts and birds, as well as visually spectacular events of note in the course of human history, the coronation of Queen Elizabeth, say.

None of this matters. I mean, the subjects of View-Master slides aren't what counts. It's the utterly elegant machinic nature of its design—Steve Jobs notwithstanding. It is the decision to work with a circle—a wheel—rather than with the planar Victorian rectangle of yore. It's the simplicity of its compactness—the poetry of there only being fourteen images for a total of seven precious views per disc. If the Victorian stereoscope resembles an opera glass, the View-Master brings to mind a welder's visor, it's true. But it's the new invention's ingenious geometry that gives the space-age artifact an aura of grace, and it's the fact of its being manually "powered" that makes for magic: the lever one presses to advance the slide is like a personal pull cord on a curtain of surprise.

What makes me feel that the "click" and "shussch" of the lever and its internal spring, the alchemy of metal against paper against plastic, is like a rope that brings a pail of water up from a well? Each Claymation dwarf is doing something precise and different in this View-Master's elaborately detailed world: one tends the oven while another inspects a piece of nibbled bread; one holds a candlestick before the gingham curtains they have sewn. A small dusting of white powder frames an elfin foot, sure sign of yesterday's snow fall, or is that flour fallen from the mixing bowl used to make the bread? Now, when I press the lever I experience the metal clasp of a childhood winter coat; now the hinge of a door, but not just any door: it's the door to a sickroom and the View-Master, my consolation.

Is it the device's affinity with a slide show that leads me here? Does the fact of its being a kind of vertical carousel evoke a merry-go-round, which evokes a panorama, which evokes a slide-show loop, which reminds me less of pictures than the warmth and whir of fans that supplies an atmosphere that weaves a blanket—is it this that makes me think a View-Master is the thing I'd most want in the hour of my sickbed need? Or is it just an excuse for a spell of solitude? The cardboard envelope into which the discs are slipped includes instructions on their maintenance: "The pictures on this reel are made on full color Kodachrome film and reasonable care should be exercised in handling. Clean with a soft cloth or camel's hair brush." I'd like someone to stroke me with a camel hair brush, but

I'm not sure what one is or where to get one; nor am I sure there's anything at all "reasonable" about dusting tiny photographs with such a brush. I only know that I'm not interested in the type of slides that read like comic strips, however painstaking their double-paned creations might be; and I'm only mildly drawn to photographs of places. It's the curiously imagined built worlds in miniature that compel me—those irreal and literal proscenia whose scale looms simultaneously intimate and vast.

Of course these things I behold are photographs, and a photographer was responsible for the fabrication of their shadows, their mysteriously suspended parts, their genies and ghosts amid the solid tables. But it's not the photographer I think of, or want to know. I think, instead, of their maker. Who was the crafter of the wave that holds the Little Mermaid? Where did she find her "stuff"? How did she choose her scenes? The sliver of triangular coral in these oceanic sets must be her signature, or the tiny conical shell the sphinx-like Sea Witch grants the mermaid (to make her human). Her head pokes through the surface—not the artist's but the mermaid's—her body is submerged, but just a bit of tail peeks through. The waves crash above them as in a painting by Hokusai as she holds the drowning prince in his androgynous swoon. How did she create the bubbles and the froth? I mean the artist not the mermaid.

Now I picture the forgotten artist/sculptor as a salesclerk at a lingerie counter by day, the maker of these puppet theaters by night. I imagine she lives on the dense outskirts of a major metropole in an apartment that we'll call a two-room flat. The kitchen table doubles as her studio, while on the daybed sleep her baby twins, the Brothers Quay. The walls are gray, and she's forbidden to paint them so she dresses herself and the boys in various shades of red. It's all ether in their rooms and painted putty, and she doesn't bother to set aside the cellophane windows she is cutting on her table, the tiny ax for a beanstalk's giant that she carves, the blueprints for simulating puffs of cloud, or securing a fake fireplace's flame when she fills their bowls with porridge.

With the invocation of any atmospheres there are adjacencies, and these are mine: I'd say, "suddenly," but it's more like "calmly," the way a memory just now inexplicably enters and then departs. It's of a scoop scraping the sides of a metal bucket nestled inside a three-wheeled cart; the memory of ice meeting air and rind meeting tongue, of Dixie cup–sized fluted containers beaded with sweat, and the grainy pungency of South Philly lemon-water ice leaving a beard to dribble down a child chin. All of this is intermingled with the scent of my Sicilian relatives' baked macaroni. The memory gathers inside this room in which I write, but it hails from other rooms where the water ice man was outside while the macaroni was cooking inside, but I could smell the baking outside and insisted on sucking the ice—no forks or spoons allowed—at the dining room table inside. I wonder if these rooms are still extant? Not the tiny row-home rooms with the dining room table that fed a crowd of twenty, but the tinier rooms of miniature verandas and voluminous velvet forms.

~~~~~~~~~~

Somewhere I had read the formulation—I think it was in an interview with Gayatri Spivak—that when we write, we are in conversation with the dead. In another place I'd read—probably in the work of the object relations psychoanalyst Christopher Bollas—that when we are in a mood, we are in conversation with some former self, the shadow of which lives on in us and is even constitutive of us, but that has never fully come into the light.

For a period of time that feels like a short time, a blip, but that has, no doubt, been a long time—certainly weeks and months, I've been having a relationship with fairy-tale View-Master scenes and with the person who created them. I've been imagining her and I've been imagining with her. Neither she nor I have had to verify the fact of each other's existence for this hazy mood realm to materialize, for the mood especially that is an aesthetic experience—and who cares if the art in question is a form of kitsch, is like a spell, and it needs no confirmation from the realms of Wikipedia to maintain its existence. In fact, the at-my-fingertips data mine that is the World Wide Web considerably complicates, stalls, interrupts, and even negates the process set in motion by the Sawyer-Gruber device. Or maybe the problem is that the storehouse that is a portal to infomercials-in-infinite-series transforms the aesthetic encounter *into* a process when, in actuality, it has no beginning and it can have no end.

In an unlocatable somewhere between the tangible and the intangible hovers the tang, the tango, the tangle that is mood. Mood moves in and passes through; sets up shop or stills us like water. It tosses us like air. It hangs like patches of color that today or tomorrow line the walls of the mind's estate. Picture our moods as our own ghostly presences in 3-D: to whom are such scenes visible? Mood as territory that can be encroached upon, or pheromone-al scents that attract some into their sphere and repel others. In my mind's eye, as I've already noted, the author of my View-Master scenes lives in a metropolis inside small rooms, an image to which I'd now add, crowded with hat boxes missing their hats. She's a person who experiences the bustle as an expanse. Unjostled by crowds, she floats. The only thing that she lets distract her from the concentrated work of affixing blindfolds to Snow White's Seven Dwarfs is the call of her calico cat to play or stroke or need to eat.

It seems I'm not the only person who has wondered about the identity of the figure who crafted the diminutive sets and even tinier puppets, their worlds flooded with a light, other-worldly. I suspect that anyone who has been graced with a View-Master of the fairy-tale type has either found themselves dreaming her into being (in which case, we try to stay in the View-Master mood) or gone in search of her (in which case, we maybe believe we can find our way back to the material world of which the View-Master is the afterimage: the shadow of a cloud).

A Google search brings her forth instantly with the header "Unsung Geniuses" in a blog by Canadian-based BK Munn, called *Mystery Hoard: Comics, Culture, and Class* where, according to Munn's information, the person who crafted the earliest View-Master miniature theater worlds starting in 1946 and continuing through the

(10. *some fine stranger no longer living but alive*)

1960s was a woman named Florence Thomas. The blog includes a black-and-white photo of Thomas poised in unsmiling profile, peering, stage left, into one of her hand-hewn interiors. Thomas's arm reaches, Alice-in-Wonderland-style, into what looks to be the combination tearoom/library of Merlin the Magician in order to affix a teacup to a table (fig. 10). She could be in her seventies here, her white hair pulled back in a bun, her face a little worn; she's wearing a baggy burlap-textured dress, maybe her artisan's smock, and I note the way she leans in, holding with her left hand onto the vertical beam of the fairy-tale wall she has fashioned, lending a new definition to the phrase, "for dear life."

The brief tribute sounds a note of appreciation for the artist whom Munn describes as having been "largely forgotten except [by] a few collectors." The collectors, perhaps like all collectors, earn a name that describes a type of person or a species apart — in their case, "3-D people," and they strike me as hailing from a place called "Geekdom." Those I find making appearances on the Internet seem vaguely "unwell." Am I one of them? Are these my people? To reverse the famous phrase, a little less than kin and more than kind, are "3-D people" motivated by a similar conditioning by mood—a lost mood, a sought-out mood, a grammar of the self's future mood?

Let's get one thing straight: I'm not a collector, or if I am, I'm the sort who's a collector-in-denial. I don't go looking for View-Master discs, and I definitely don't aim to amass them; I only need a few, preferably happened upon at a yard sale. For me, the calculus of seeking and finding has to retain an element of accident. Using the Internet to "look for" View-Master discs entirely misses the point for me, and I can't think of anything more likely to negate my relation to the object of pleasure. My form of collecting might be much more delusional, less transparent, laced more with the privacy of pathos than that of he who seeks and finds, and finds again because I think what I'm after is a biding of happenstance, and the kernel of the quest is of the View-Master finding *me*.

YouTube videos give me access to the woman said to have amassed the largest number of View-Master discs in the world. For a total of thirty thousand reels, it could be a Guinness book world record: her reason for collecting, "to get away from life's other problems," but the word "other" suggests that her collection is, in itself, a problem, even if a more manageable one. A man who identifies himself as J. Clement, who is in the business of restoring View-Master reels and who expresses a partialness for the work of Florence Thomas, which he understands as unsurpassed, delivers many out-of-print scenes that far outstrip my small collection: apparently Thomas dabbled in sci-fi fantasia like *Sam Sawyer*, and *Tom Corbett, Space Cadet*; she made an illustrated untitled alphabet—imagine learning one's letters through her mood-imbued views. He names other producers of figures and their scenes who either served as Florence's apprentices or who worked alongside her but never, to his mind, equaled her— people like Joe Liptak; Leila Heath; and Frank Visage. No one quite achieves the detail of Thomas's dioramas and panoramas, her sense of background or depth. Waxing dreamily on her rendition of *The Sword in the Stone*, Clement fantasizes a kind of personalized *if only*: imagine if she'd tried her hand at his *own* favorite tale, he asks, *Mistress Masham's Repose*, also by T. H. White. The sculptures are gone, he says, because they weren't properly valued, but at least most of her creations' slides have kept their color: Kodachrome retains its vividness seventy years later, but most post-1976 film stock will fade to magenta in twenty to forty years—a monochrome mood forever afterward cast.

Clement's uploads are informative but frustrating, even oddly sacrilegious, because isn't the mood of Thomas's creations dependent on the visor effect? On climbing inside the black box with her? The mask as extension of our face and our eyes? Why Clement translated the View-Master into a 2-D YouTube slide show is beguiling— but that's assuming I can actually view the out-of-print Thomas memorabilia he's amassed. Without a 3-D viewing apparatus built into my computer, I'm left to strain and squint at red-green transparencies that don't line up and can't even hint a figure into view.

If I want to go further on an Internet quest for Florence Thomas, I have to purchase something. An as-seen-on-TV website offers a DVD with what feels like undocumented pirated footage of TV shows on which View-Masters appear from *The Mary Tyler Moore Show* to *The Simsons* [*sic*], from *Bold Journey* to *Good Eats*, from Conan O'Brien and Letterman to Colbert. One episode of the 1950s *You Asked for It* (also known as *The Art Baker Show*) promises to answer the question of the little girl who inquired how they make View-Master scenes "look so real"? In place of the singular Florence Thomas, a whole team of creators appears that includes Mr. Lee Green, artistic director; Walter Chaney, sculptor; the suddenly gendered, wardrobe *mistress*, Martha Armstrong; and, the incredibly named, Blanding Sloan, scene designer. In a wonderful moment, Sloan is shown bent over a fishless aquarium. With no actual fish to feed, he drops bits of diamond sand through a sifter—could it be sugar?—to create a gleaming ocean floor. In another scene, Armstrong points her awl at the accurately diminutive gold braid she has sewn on Captain

Nemo's sleeve—"it is militarily correct." "He wears a silicone collar," our narrator explains, and, beneath his jacket, an entire shirt.

The show has the feel of giving us an inside view of Santa's workshop, but its emphasis on authenticity seems to miss the point because the thing about this genre of View-Master, when it comes down to it, is that, rather than assail us with information, it beckons us with dream. If it is painstaking in its assemblage, that attention to detail moves us to *lean in*. Outside switches places with inside and vice versa in this exchange: View-Master dioramas don't look "real"; they look other (worldly): that's their appeal—their success in having molded a portal that crosses the far-off with the known. Once lured into their irreal precincts, do we wish to flee back home or seek to stay? By imparting something of the dreamy and inviting us to enter in, perhaps they pay homage to *our own authenticity*—the real that lay buried in a psyche as sea, the authenticity of an inner life.

<center>~~~~~~~~~</center>

Human concourse requires that we share the same space and adjust to an agreed upon dimensionality. Instability is often understood as a door that's come unhinged. Enter View-Masters, dreams, and moods: only in those realms, do we get to alter the terms of that relation. What's weird and wonderful about Florence Thomas's fairy-tale tableaux is how they play with thresholds and alter our established modes of egress and access. Is all mood a species of reverie? View-Masters seem to court our capacity for such. At the very least—but their least is vastly bounded—they resupply a form of trance to every *entrance*, translating every doorway into a form of suspended consciousness, every foyer into a place of untried mood.

There's a school of psychoanalysts that treats the ego as pellucid or skin with porous boundaries. You can imagine the fun they have with metaphor as the ego becomes ever more abstract: now they call upon a Möbius strip, now a "non-orientable manifold," now an envelope to picture the somehow stabilizing seat of will, emotion, and mood. After reading articles in this literature, I'm left to pic-

ture my Self as a type of filmy origami. In one such article, a woman
is described as having the following "transferential phantasy": that
"during her sessions, [the analyst] left by a concealed door at the
back of the room, behind her, in order to reappear threateningly in
the picture that hung on the wall in front of her."

View-Master to hand, I go in through a vent to arrive at a sofa;
in through large Alices suddenly reduced to small. I go in through a
tiny bulkhead and out through a tree limb. Out through a book and
in through a map. Out through a curtain and in through a cloud. In
through a toy chest and out through a pantry. In through a hallway
and out through a cocktail party of playing cards. I go in through a
lamp and out through a doorknob. Out through a balcony and in
through a pyramid; out through a thatch and in through a cobble.
In through a hole in the ice and out through a marsh. In through a
sliver and out through a fold. In through a carriage door and out the
vaulted inside of a walnut.

For another group of psychoanalysts—those of the object relations school—adult moods almost always have embedded within
them some form of *past-ness*. Whether they be representative of our
own past states of mind, our conception of ourselves in the past,
or re-creations or reexperiences of former infant-child states, they
stand to do their most generative work when they put us in touch
with what psychoanalyst Christopher Bollas compellingly calls the
"unthought known." We are the recipients of enigmatic, which is to
say, untranslatable and eventually repressed because inassimilable,
messages from birth, even simply by the touch of our earliest caretakers. Such "untranslated residues," psychoanalytically speaking,
are constitutive of consciousness even as they remain out of view.
It's the perennial conundrum of our humanness that is at issue—
the fact that we spend a good deal of our early development without
words even as we come to equate consciousness, and with it, being,
identity, and subjecthood, with our language-making capacity.

"The subject arrives on the scene rather late in the day." Bollas
only seems cavalier in his condensed account of this fundamentally
belated relation to selfhood we all share, for the accumulation of

wordless impressions forms the bedrock, in his view, of the spaces of later seriousness or play, unbidden gloominess or recalcitrant colorations that are our moods: "By the time we are capable of a meaningful interpretation of our existence, and the meaningful presence of others, we have already been constituted via the ego's negotiation with the environment." Whether he terms this unconsciously stowed "stuff" as the "shadow of an object," "untransformed being states," or the unthought known, Bollas understands it to resurface in mood states, and to do so conservatively, generatively, or malignantly. Being present to another person's mood state is, according to Bollas, a kind of sacred share, and one perhaps not fully possible outside the contract struck between psychoanalyst and analysand, not least because to be in a mood is to broach realms of wordlessness wherein lie our (presumably) truest selves.

Moods, in this light, are the things we turn to in order to express what we cannot narrate and have no words for. Ineluctably private, moods in a sense *manifest* some persistent remnant of experiences that exceeded our capacity for representation or symbolization. Inside our moods, we are put "in contact with that child self who endured and stowed the unrepresentable aspects of life-experience."

Of course, Bollas is not addressing mood as equivalent to our *ever-present* field of consciousness as such; he's more interested in moods as special states or forms of *altered* consciousness, enabling (or disabling) of psychotherapeutic work. When he asks, "How far inside the mood is someone? How long will it last?" he hints at a kind of sacredness to mood or temporarily autistic zone or "spell," that, as he puts it, "is often licensed by a recognition of its necessity." Sometimes I picture him waiting for a patient's mood like an audience at a magic show expecting to be startled; other times, I find his questions moving because so genuinely interested: "Who is it," he asks, "that emerges from within the mood?" And here he's following the cue of fellow analyst Paula Heimann, who he tells us was among the first to resist a more classic view of the analysand as a neutral reporter of his inner states. Heimann proposed instead that analysts acknowledge how at "any one moment in a session a patient could be speaking with the voice of the mother, or the mood of the father, or some fragmented voice of a child self either lived or withheld from life."

Moods can be the placeholders of some hardened, frozen, arrested, cutoff piece of experience—in which case the person holds fast to the unthought known, conserving it nearly as a personality trait (or rigidly defining mood), rather than finding a way to bring it into language, and so, to transform it. Here, in Bollas's view, the earliest caretakers are to blame for their inability to tolerate, respond to, recognize or even acknowledge some piece of experience brought to them affectively or somatically by their infant or child.

If our environment, not just our parents, cannot withstand us, what do we have at our disposal for producing those environments we come to make for ourselves: our moods? Absence plays around the edges of our moods: maybe our moods are our means of making something out of nothing, or is it the other way around: they are our way of making nothing out of something? Sometimes our moods require that we tune others entirely out; other times, we can be partially absent by way of a mood and still be present to others. Even if moods aren't attention-getting devices—more likely excuses to be left alone—the question remains of the allowances they require from the world or from others. View-Masters might just be a means of gaining access to moods otherwise unallowed. No one ever said, though I suspect it's true, that they might be fundamentally antisocial.

~~~~~~~~~~

A tower of playing cards rises to the height of two compactly molded spruce trees. A five of diamonds and ace of clubs bend outward from a show of clouds and mountains whose register in a further distance seems held in place by a copse of trees in a layered middle ground. One frog, one lizard, one mouse, one white rabbit scatter, but Florence Thomas's Alice in her aproned dress tilts with the cards, half-scared, half-intrigued by the potential havoc their toppling will wreak. Alice's head is suddenly too big for her body—is she in a toy store or a courtroom when she knocks over a box filled with woodland creatures as plentiful as Noah's ark? A tortoise waffles, a brown squirrel prepares to leap, while Alice's startle is centered by a table spread with barely touched tarts. In another scene, a collared piglet accompanies Alice outside a cottage whose shutters and clapboards could come loose if a wind were to enter into the scene, like a nail stuck to a finger by a shred of skin, loosened by a childhood accident of shuffle board. Someone didn't see me crouching—and what I was doing, as I remember it, was trying to determine my own size and place in the world, when, bent all the way to the rubbery surface of a shuffle-board court, I pointed my index finger to the ground to feel how high the disc was: it was a game without an intended answer, more an experiment in feeling the difference between skin and marbleized substance, a lozenge too big to ingest but just the shape of a candy Life Saver that was part of an oversized playing board, until someone, not seeing me, jettisoned a disc in my direction, smashing my nail to purple until it eventually fell off.

Inside a Florence Thomas View-Master, is that a fabricated fountain or a quiver of glass? If she used merely a twisted pipe cleaner, how did she achieve the upward movement, the pulse, the flow? A duck's nest might be manufactured from a peach pit. Objects float without receding. Tree trunks aren't simply brown but variegated shades of brown; globule and animate, their leaves like bunches of curls. Playing in diminutive space, Florence Thomas has no room for shortcuts: she sets the Mad Hatter's tea table with not just one or two or three tea cups but ten, hand-crafted, each the same and different. There's a magisterial mansion shrunk to bite-sized. And I find beauty there too, the delicacy of a duck's reflection—but it's "The Ugly Duckling" she is illustrating—sheathed by lily pads. Not exactly frolicking with Florence Thomas, it's more like I am nudged by her toward awe, just now drawn back by tears withheld from sadness. Where was I when I was held at a table of childhood cards but the adults were there too, insistent to know what I had in my hand,

and I liked this feeling of playing together in the round or in the square, but I also maintained the desire to flee, to run, to become a card, and fold up like the table.

~~~~~~~~~

"Can you remember when you used to look through these?" a ghosted, barely audible off-camera voice asks. And then replies, "Well, I'm happy to say I can't quite remember when I used to look through them . . . this is an old stereopticon-type thing."

Rather than train itself on the object the voice refers to, the camera instead pans across a phrase that is embossed on the side of the TV studio's cameras: "color television." If the *Home Show* had been in color, I'm missing it since the as-seen-on-TV DVD footage remains in black and white. After only three viewings of the DVD at nineteen dollars and ninety-nine cents, I can't even revisit the footage if I wish to study it because I begin to get a message from my computer that says, "Skipping Over Damaged Areas." Settling for this form of bargain-basement history, I remain steadfast and forgiving: this was live TV so there was no chance to edit out mistakes, and the program promises a glimpse of Florence Thomas.

Our host, a heavily made up, pearly-necklaced, cinched-at-the-waist-maybe-housewife explains that we are going to take a look at a modern version of the stereopticon, and better: we are going to meet the person behind them all, "the person who created the idea of their presentation in a new form," she whose forms capture something with great delicacy, and who has studied "sculpturing" both here and abroad, "Miss Florence Thomas of Portland, Oregon."

"Hello Mrs. Thomas! . . . *Miss* Thomas, I mean," begins the first of the host's many flub ups, or "bloopers."

"Why, hello Arlene," Florence Thomas fumbles a bit with her microphone even though it is slung securely around her neck. Throughout the interview, she glues her left hand to her waist just beneath the microphone as if she were experiencing the mike as an appendage that she wishes or expects to fall off. She wears a dress, though I'm not sure she is at home in a dress, and has donned jewelry for the occasion as well as painted her nails. Her dress is complexly patterned in abstract leaves and flowers—my guess, in shades of yellow, green, and black—and she has the sort of 1950s hairstyle that makes it seem as though a beanie-shaped hat has left its mark in place. I imagine this show to be lacking in both makeup artist and green room so that both women were left to their own devices in making themselves camera-ready. Florence's eyebrows have the look of press-on tabs or the sort drawn in by an animator. Indeed, there is something of the cartoonish about her here: she has that quality of enthusiasm that one associates with Shirley Temple, and a grin as wide and untrustworthy as the Cheshire Cat. She would roll her eyes if she could, and offer phrases like, "By golly, by gum!" but the off-camera lights are evidently so fierce as to be blinding, until her awkwardness—a mode that seems natural with her—appears in competition with a species of tortured squinting.

In spite of her host's rapid fire, prefab questions, Florence Thomas manages to slip in answers that are novel and startling:

HOST: I wonder who your favorite character is, Miss Thomas?
FT: Usually it's the one just in the future.
HOST: Your enthusiasm is for the next one.
FT: But of all of them I believe I like Alice in Wonderland the best.
HOST: I guess most children love Alice in Wonderland the best too, Miss Thomas, don't they?
FT: Oh, also adults, I think.

When the host notes that she must have to do a lot of research to get her fairy-tale characters' costumes right, Thomas chortles, well, yes, either that or "*invent* a period for them." Where Thomas is clearly a homegrown original, too down-to-earth to hide her Western accent—pronouncing "pleasure" as "play-zhur"—our host is burdened by an affected pronunciation associated with the first talkies even though it's the 1950s, a sort of quasi-British upper-class Bette Davis-ism that requires that each word ending in "-er" be pronounced as an "ah," so that "a poor Miller" becomes "a poor Millah"; "a daughter" becomes "a daughtah"; "important" becomes "impahtant"; and "it couldn't be simpler" becomes "it couldn't be simplah."

The initial informational exchange between Thomas and her host is followed by a folly of an interlude: the TV producers have placed an array of Thomas's figures and their sculpted settings entirely denuded of their 3-D surround, widely varying in scale, on a rotating surface, which the camera shakily pans as prelude to the host's "reading out" "Rumpelstiltskin" while the 2-D camera moves from scene to scene. "Pretend to look through a View-Master," we're instructed, but this setup infers nothing of the sort. To make things worse, the host tells the tale in Martian voices, and there is also a musical soundtrack reminiscent of ads I recall from the era for the *Longines Symphonette.*

Our host's performance closes with a brief exchange of compliments between Florence and herself as they ready themselves to end the show until Florence, fumbling behind her, catapults to the rotating table to grab something she intends to give to her. "Uh oh!" the host exclaims, as though talking to a Teletubby, "What is that?" Was the audience for *The Home Show*, also known as *The Stay at Home Show*, children or housewives, both, or housewives understood to be children? Why does the host think that Florence Thomas, inventor of photographic methods for documenting dioramas in 3-D and consummate creator of same, need to be addressed condescendingly, as a child? Did her reciting the tale put our host into a childlike trance?

"I'd like to present this to you, as, eh, a memory of the happiness we had, eh, ah, with your visit to Portland," Thomas recites her unrehearsed extension of gratitude, and hands our host a figurine she has fashioned of . . . the host herself!

"How absolutely wonderful," the host blubbers. "Well, I'm no fairy tale either, and there's my diamond heart! It's *just* lovely. This is *all* for me?" she says as if she has forgotten to count because babies can't count. Beneath the exuberant flood of baby talk, Florence Thomas replies as though smitten with the host, "I think you are," as in I think you *are* a fairy tale. You live inside a TV just like I live inside a View-Master, she might have said. Or maybe she just means such an overly performed figure cannot possibly be real.

"That *is* for you," Florence goes on, continuing the game of feigned enthusiasm, and the host concludes with a mistake-riddled speech: "Well I don't need a reminder of the lovely time we all had in Portland, Maine, Miss, ah, Portland, Oregon, rather, but if I needed any reminder this is a very nice one."

The program closes with an ad for View-Master during which the host, once again, reads the story while the children, visors pointing upward toward the light, watch.

"Everything looks so real!" she marvels, "It's like being in the story myself."

Some things illuminate by negation, and this show is one. View-Masters, I now understand, aim to be wordless activities. If someone were to read to a child while he was enclosed inside a View-Master, the child would be justified in hitting that person over the head with the device. I even doubt that any child ever reads the frames silently between each view. The magic of the View-Master isn't that it brings the story to life, but that it takes a child, or an adult, to a semblance of life *before* narration was possible, the space attuned to the sounds that predate language as precursor to our moods. Peering through a View-Master, we might feel as though "we're alive inside the story," but establishing points of identification isn't its game. To help us to feel alive inside a sphere that is decidedly not lifelike. That might be more to the point; that might be its power. Partly there, partly not there: that's the nature of the lively nebulae that is mood.

A few more strokes of an Internet search and she's there: not Florence Thomas, but *The Home Show*'s host. She's Arlene Francis, whose later claim to fame was her role on shows like, *What's My Line*. *The Home Show*, I learn, was part of a talk show trilogy that also included *The Today Show* and *The Tonight Show*. Francis could have gone far as the first female talk show host in television history according to one Internet source on the subject, but a combination of her own internalized sexism and the sexism of the industry held her back. Some episodes of *The Home Show* are viewable on YouTube. In one, she is interviewing Jackie Kennedy at the time when John was still a senator. Together, they view a home movie that follows Jackie as she runs her daily errands in Georgetown. The sexism that Francis's questions perform is so tacit that it makes the show too painful to watch, at one point using the time with Jackie to glean an apparently important piece of information, the sort of data that viewers would want to know: without blinking, she asks if Jackie ever irons John's pants. In another episode, she interviews Charles Eames of furniture fame with the announcement that behind every great man lurks a woman. Accomplished artist, Ray Eames, appears as well, and seems a little taken aback by Francis's positioning her as second fiddle to her husband's first.

Behind Florence Thomas, apparently, no one. Florence Thomas is ever out of view. There's hardly any trace of her on the Internet, and I've yet to discover even a birth date or a death date. Though I've purchased an in-house book compiled by collectors that promises to offer a little something more of her, it's clear Miss Thomas left no heirs; she had no progeny but us. Better that she not be fixed, filled in, but remain a figment, and a fragment, to a world of lucky strangers, child and adult, hopeful to receive and then accept her invitation to entertain and open toward, to roam around inside and portal through, to ensconce inside the nonsense of a mood.

synesthesia for orphaned boys

Wire the tail to the body. Tie the feet together with the data la-
bel string. Adjust the plumage, and stuff the eye sockets and throat
with cotton. Tie the bill shut, and rest the bird on its back in a
cured piece of wire-cloth to dry. —Leon L. Pray, *Taxidermy*

The phrase "I'm in the mood for" usually hovers somewhere between a preference and a craving. At any rate, there's movement in
it, and promise. Even if its completion is part of a repertoire of repetitions, it still holds out the possibility of our being capable of something new. "I'm in no mood for" seems so much more final a phrase,
and I always imagine it as spoken by a parental figure to a child,
particularly an adolescent, and culminating with "your guff." If the
teenager being addressed were female, the sentence would end with
the words, "young lady," as crowning bonbon on a layer cake of condescension passing for an appointment to a higher station. I can
think of worse ways of being addressed than "young lady," but not
too many: for pairing diminution with negation, it pretty much takes
the prize; for erasing a girl child in the name of putting her in her
place. I should therefore consider myself lucky that my own father
would never have addressed me as one capable of guff; nor would
he have ever called me "young lady," though I do recall his pointing
to photos he had set aside from a family outing of my mother and
me. I didn't like that he had hung the photos like talisman or voodoo
dolls from a contraption in his bedroom, part burnished metal, part

naugahyde, called his valet. He'd point to the photo occasionally and say, "dem's my gals."

"Guff" would have been too refined a word for my father, but that's not the only reason why he would not have addressed me this way. Quite simply, he rarely addressed me. Being in relation to my father entailed tiptoeing around the zones he inhabited, or unconsciously trespassing their borders—which was easy to do in the row home we shared and given that most of his mood work seemed to be carried out in the living room where he sat solo watching a TV program that only he enjoyed. Since the front door of our house entered directly into the living room, there was no getting around the need to cross the barrier—more like a force field of trance—that defined the space between my father's nail-biting gaze and an episode of *Wild Kingdom*, a program that tracked, in safari-like fashion, the strangely tantalizing world of our animal brethren. I might be attempting to cross the line with my black-striped track shoes, fresh from a meet and trailing gravel from a scene at which he never made himself present as witness or coach; I might be crossing with the impossible-to-mute slapping of a flip-flop; the thud of an oxford-tied school uniform shoe; the buoyant surprise of an Earth shoe, but the scene on TV was ever a variation on a theme of evisceration, a lion going in for the kill, a cobra digging its venom in, the preternaturally languid pulse of something bloodthirsty and predatory. Even if the background was pacific as a silent stream, eventually there would be pursuit, the lying in wait followed by the pounce, the shudder, the twist and heave: from the corner of the eye of my interruption, I'd catch a whiff of entrails.

Like an animal, my father had a good deal of bristle about him. Mood-wise, he was bristly. Crude. You add an *r* to "guff" and you get my father: "gruff"; you alter the letters slightly, and you get the sound a dog makes, "grrr—ruff-ruff." Was my father working out the terms of his own animalistic rage in his Sunday afternoons with *Wild Kingdom*, sponsored, ironically, by the Mutual of Omaha Life Insurance company? Was he trying to learn how to live with the animal within, and is this why I never gained the slightest degree of comfort in the wild, whether it be a walk in the creaturely woods or a domestic cuddling with a furry-eared friend? My father didn't allow my brothers and me to have pets—there were to be no creatures that might avert his gaze, nothing that required care beyond a modicum of watching and watering, no teeth-bearing Other padding the stairs in the dark—but this doesn't entirely explain my own adult fear of animals, my conviction that I am entirely without knowledge of how to "deal with," "relate to," "handle," or be with an animal. All of that is just an alibi for the depth of a childhood spent as witness and benefactor of a kind of human wildness sadly confused with "animal": my father's fundamental inarticulateness and liability to explode, lash out, strike, violate—and let's not forget that he ate with his mouth open.

My partner, Jean, helped me to live with the animals, coaxing me to hold the harmlessness of a mouse in a jar if not my hand; insisting I go to meet the neighbor dogs so as to tame my fear of them; offering that old saw about the bat being more afraid of me than vice versa, or the bee having simply confused me with a flower, even helping me to discover that I was capable of falling in love with a cat to the point of stroking, nuzzling, and kissing her on the top of her head, feeling her lick in turn, but never to the point of allowing myself the luxury of picking her up, convinced that she'd scratch my eyes out if I were to try.

That's the one thing that can't be taxidermically preserved—an animal's eyes—so they're forced to fashion expression out of glass, the taxidermists are. I know this from having read many books on the subject during a period of trying to write a book on the subject. I had gotten as far as devising a clever title—"Posthumous Postures," I explained in a letter to a friend, offering as précis that it would be a meditation on scared states, wild states, and skin, with trips through and around Hitchcock, Thoreau, the ornithologist/photographer Cordelia Stanwood, Raphael Peale, and my own psyche. I was intrigued by the language of taxidermy: the implication of words like "freeing" and "relaxing" skin; the desire among practitioners to distinguish between "mounting" and "stuffing"; the taxidermist's need to wrestle with unresistant matter. I wanted to understand the nature of the dread that might lie behind some forms of taxidermic practice; what it means to be inside *human* skin, or whether we humans ever assume our natural form. I wondered if, by some type of inverted reversion, our most natural form might be the pose we struck when we manifested fear of or aversion toward the natural world rather than a kinship with it. I was taken by sentences I couldn't possibly make up, like: "An old squirrel has a skin of Herculean strength that makes a satisfying den trophy"; or, "There is real satisfaction to be found in collecting, mounting and arranging a personal collection of animal specimens. . . . Study this book, then try your hand, and know the thrill that comes from doing work which at the same time can be stimulating recreation." I had a hunch that taxidermy is really not so simple or easily summarized as a practice or as an art. Nor as an attitude, or as a tradition. Nor as a pastime, or as an aesthetic. Nor as a process, or a set of acts. An impulse, a need, or an unthinkable arena of human willfulness, it requires care, planning, and a conclusion, set in feathers (if not in stone).

My own conclusion was that taxidermy makes animals into idiots and was therefore a form of reckoning, not with death, or humans' obvious dread thereof, but with the death-in-life that was human stupidity. Taxidermists were modern folk caught in a stupor of having forgotten how to live.

And what did I know about living, say, in a cabin in the woods—an adventure tinged with the ethereally calming scent of baked pine needles, Jean assured me, when I agreed to purchase a log cabin with her perched on a steeply graded granite ledge in Downeast Maine. "I can't do it! I'm not going to be able to live here!" I literally cried, retreating to a corner while the mice lorded over me as they waltzed across the kitchen counter where I chopped the garlic for the sauce, content in their slink and scurry of a claim to the kitchen as home. It's *just* a mouse, *just* a flying squirrel, it's just a blind rodent with paddles for paws, in other words, it's *just* a vole, Jean would say, when a "critter" would appear in the house, dead or alive.

Jean was capable of touching animals in all their multi-formity, alive or dead, sleeping or awake. She once gently lifted a live bat from its scabrous scaffold of bricks in our Providence home, and when it came down to it, she had no qualms about lifting a gray-blue short-haired rodent, tail and all, from its death pool. It was the sight of the squirrel floating in the toilet's anti-freeze that greeted us on move-in day that brought out the clichéd cottager in me and haunted me to the point of tears: signs of the squirrel's entrapment were every-where, from teeth marks on chair rungs it might have been trying to eat, to the shredding of window panes it had tried and failed to es-cape through. If I felt such compassion for the squirrel, why couldn't I get close enough to aid Jean in lifting its carcass out of the toilet bowl where it had died, or at least agreeing to hold the bag for its disposal? What mechanisms of idiocy, or of ownership, were re-quired to convince me that there must be someone skilled in lifting bodies of animals out of commodes who I could hire to do the job? For my own part, I felt it wasn't something I could do without in-structions, methods, tools, maybe even some form of counseling, but we did it, and buried it in the yard, but not before the thought of taxi-dermy as a form of symbolic preservation entered my mind, a species of possession and possessed, surprising me enough to want to try to understand the singularly human practice of remaking dead animals into the mold of our own likeness.

I never finished writing the book; nor did I mourn it when life— some would say, accident, others would say misfortune, still others would say disease—intervened. All writers have books and essays and stories and poems that exist as husks or nest straw or moltings that do the soul's bidding in the dark and don't ask for the light of day. Had taxidermy ever really been my subject anyway?

It only takes a rekindling of thematic twigs to find the fuel that drives the flame of our interests, the surround everywhere apparent and invisible: the campfire's mood; the angler's ambiance; for me, the penchant for dioramic picture windows inside of which taxi-dermy was just one contributing detail—a fascination for natural history museums dating back to my having been raised in proximity to the city in which natural history was birthed in this country by Charles Wilson Peale, the town of Darby, named for a town in England, tonally registered as Quaker though populated in the 1960s by Baptists and Catholics, Episcopalians and Jews. No: dating back to eating sundaes at our town's Woolworth's lunch counter, its mood of seediness and grilled cheese. You could be treated to a hot dog and twirl on the seat while you waited, or try to fathom the place of live birds—usually yellow-feathered or deep blue—chirping in short cages amid bottle openers and Dixie cups, T-shirt bins and toilet paper rolls, hammers and party ribbon, dental floss and step stools, candy and gift cards, oddest of all, amid crates of bargain books going for ten cents apiece. The "pet" section smelled, as did the lunch counter, and in neither case was it a good smell, but there was an incitement to roam and wonder and be afraid and delighted and disgusted all at once that maybe marked Woolworth's as my earliest experience of a museum of natural history.

Our lives are sedimented: every locus of our present being just one shelf inside a layer of otherwise invisible shelves nestled like Chinese boxes. The older we get, the deeper the cabinetry's plumb line, the more rickety the expanse of, just now, this study in which I write, a desk, inside of which resides an earlier instantiation of the writer or reader at work at a dining room table, and, before that, a metal-topped surface on which my father had us lay pinecones in a rare moment of serene sharing, and before that a Catholic schoolroom desk with defunct ink wells inside of which I imagined monsters, and before that a play table shared with fellow kindergartners, its underside lined with boogers and gum.

Enter a natural history museum and aspire to a condition called "debritude"; experience the lively bric-a-brac, nature's flea market that hints at the soul of things and hallows mere rooms into temples. Once inside a contemporary museum, do I comingle with the Woolworth's birds of yore, rubbing necks against their feathers? And what does it mean that a mood of intimacy is made possible by the absence of the live creature, the animal made fully present now, hushed but proximate, held inside an architecture, perched inside a room whose display is dependent upon its death?

~~~~~~~~~

I have a friend who remains an avid reader of newspapers, who scouts out articles of interest like everyone's grandmother once did, clips and sends them to me in the mail. Caren is the sort of person who listens for what moves people and then tries to help them find it. She keeps on the lookout or goes digging for any relevant material, like a research assistant, to the work of your heart. When we are nearer to her, in summertime in Maine where she lives year round, her usual greeting is a menu she's prepared for us of off-the-beaten path excursions.

One year Caren told me she had just the place for me, but she couldn't describe it, we just had to go and trust it would be more than worth the ninety minutes it took to get there. "If it's taxidermy you're interested in, this is your place." And that's all that she was willing to say. The place was a natural history collection dating to the 1920s, the L. C. Bates Museum, located on a campus of a one-time orphanage of sorts in Fairfield, Maine, named Good Will-Hinckley, founded in the late nineteenth century as the Good Will Home Association, by a god-fearing philanthropist, Reverend George W. Hinckley (1853–1950), a.k.a. "Gee Double You," as part of Hinckley's quest to educate and shelter, at first, needy boys, and eventually, girls as well.

I doubt that anyone on a first visit to the L. C. Bates Museum could assimilate the gestalt that somehow yokes the atmosphere of the museum with the ephemera documenting Hinckley's mission there—"Living the Good Will Idea"—mixed together as it is with artifacts so vastly varying in scope and so kookily juxtaposed as to daunt a visitor to its precincts with delight and wonder. One minute you are being watched by the black eye of the detached head of a white stag like the vigilant gaze of a trick painting belonging to the Adams family and the next minute you are asked to submit to the title of one of Hinckley's guides for young boys, *Chore Doing,* this one is called. Darkly wood-framed vitrines host fragments of the first transatlantic cable alongside a collection of Orientalist esoterica and Russian drinking horns, which share a room with Reverend Maurice Allure's circus-model miniatures, one cuneiform tablet, a cast iron mortar and pestle, and Native American baskets woven into the shapes of fruit. There are ivory busts of the world famous (William Shakespeare) and the stuffed heads of the locally revered (Barney the Bison; figs. 11 and 12).

Atop a vertical glass display case sits the largest pinecone I've ever seen, and next to that a very large seashell, or is it a species of gargantuan seedpod, and next to that, a can of Raid. All of this is held together with an electrical system that runs on exposed binary tracks stapled to the same sorts of wood mounts that hold the moose heads in place but here, there, and everywhere grounded by, what also could count as collectible since it's clearly antique, porcelain.

The world is out of joint and yet all the world's conjoined, the carefully documented mishmash seems to say, and my first visit with

(11. *perchance to dream our way into a mood of our own making*)

(12. *unpredictable moods, their wild ways*)

Caren was marked mostly by a mood of madcap. Not implicating ourselves in any of it at first, we let ourselves be entertained by the wackiness of the museum's arrangement.

What makes decrepitude effect a mood of ticklishness or buffoonery, because aren't we also made of dust?

We stood breathless with laughter before the riddle of an elephant stump overflowing with excelsior bereft of the taxidermied body-entire from which it seemed to have been ripped before being placed on a shelf in this museum, and for what purpose? We were like unruly kids snorting soda pop through a straw in a movie theater in a room hung with poorly conserved fish carcasses. Parts of the fish mounts were eaten away—is that what the Raid had been for?— and their scales flaking off like snow, like ash. No aqueous habitat in sight, the clearly dead fish display struck a note of high comedy, but we had elbowed each other to be quiet: a couple beside us had made a pilgrimage to the room and to one of the fish. A decaying marlin had enjoyed the privilege of having been hunted by Ernest Hemingway himself, and these people were his relatives (plate 2).

I could argue that taxidermy isn't about averting decay—it is about stylizing it—but it only takes a specimen's falling apart before one's eyes to point up the foible at the heart of preservation. Decomposition can inspire irreverence especially when it pokes through human attempts at shiny spectacle. Taxidermists come in many stripes (LOL), but so long as it is dead matter they are handling, they in some way walk a tightrope of the reverent or the ir-, of honing or defilement of a once-living animal's sloughed off, shucked off, pried loose reversion to a mask. If the mask hides a face, can you find it? Do you dare to reinspirit it?

There is a mood of reverence that is the result of decorum, protocol, convention—we know it when we find it in a church, a place of worship, or of patriotic observance; it's readymade, ritualized, and it depends upon our being trained to recognize it. But there is also a mood of reverence that is the uncanny by-product of the sublime. This sort, I believe, is much harder to come by, and it's capable of catapulting us to the sort of place that privileges slumber over sleep, sauntering over walking, quiet over silence, balm over cure, weeping over crying, deep shade over shadow, solitude over aloneness.

Call me bipolar: if, at the beginning of our visit, I was laughing, soon I was filled with the welling up of tears. It wasn't as if someone had rapped our hands to quiet us or asked us to stop acting like children. It was more like some ever-hovering ghostly guide tolerated us, knowing that it was just a matter of time before we would cross a threshold like a climb into a mysterious canoe. Then a voice would take my hand and lead me in saying, "Be still now and grow into this becoming-child mood. Be kind around it and do not stir; let its sudden gradualness overwhelm you."

~~~~~~~~~~

The layout of the building is arranged in such a way to allow you to stand above and look down into a room of habitat dioramas featuring mammals—there was bobcat, marten, mink and squirrel, raccoon, red fox, woodchuck, and skunk. There was caribou. You can stand on a kind of balustrade and look down into the passing panorama, eventually to be seduced enough to descend the stairs and get a closer look. The cases run in a continuous series around the room starting in a corner with a scene of great whimsy: someone had fashioned part of a barn inside the diorama with a lacy nest hung inside an inner window ripe with eggs, then positioned two skunks below en route to rob it. The room is crammed to the point of overload with numerous unhabited mammals sitting atop these more specialized scenes (plate 3).

The lights that lit the dioramas in some cases hung down like microphones not meant to show on a movie set, and it was hard to look past two poles that had been bolted front and center to the ground to prop in place the now-sagging belly of the massive, lanky caribou. But what was different here was the presence of a vision and the touches of a hand: most of the mammal dioramas were backdropped by impressionistically styled paintings that I couldn't help but enter and be silenced by: someone had created habitat dioramas unlike any I had ever seen; this someone had captured a species of beauty.

A bobcat crouches naturalistically enough, but it's the painter's brush that touches me more amply, volubly, fully. The artist has painted in a piece of fallen cloud to match the creature's mottled coat, or maybe a cloud's reflection on the surface of a brook that runs across the same forest rocks the bobcat straddles while it prowls. A river otter's coat is the color of twigs and sleek as scudding reflections—the scene is wet; its belly must be as small as a ball just as trees are bunched fists when glimpsed from a distance (fig. 13). The center of a diorama sports a 3-D squirrel, or perhaps it's painted in? The main thing is how the painter achieves the same sense of achromatic tonalities with the kind of values one appreciates in a painting by Winslow Homer: cranberry, umber, not-quite-copper, faded yellow, cadmium white: even dried remains of plants are live upon a scene of snow, the stemming quiet (fig. 14).

Are these habitat dioramas in the traditional sense, or are they paintings with 3-D elements that have become part of a deeply felt atmosphere—a state of mind—into which these elements vanish, appear, and vanish again? In these rooms I felt that defining moment of a mood: mood as a baseline state of being that moves persistently and with radiant evanescence—even if the mood be dark—from nothing to something to nothing again.

If natural history museum dioramas are known to work against

(13. *that warming space between an outer and inner carapace*)

(14. *be kind around it and do not stir*)

the presence of the artist who constructed them, these announce a
presence, if not an ego signature: they were all constructed by the
same person, according to the placard, Charles D. Hubbard (1876–
1951), apparently forgotten American impressionist, close friend of
G. W. Hinckley, with whom he worked hand in hand in conceiving
of and creating the museum.

Does transformation require a degree of transfixation in we humans? In order for one part of a self to move across, another part
that is ever active, possibly running, must stand still. In order to see,
we must be made to stop looking. Crossing the threshold into a room
of nothing but bird dioramas by Charles D. Hubbard, we are graced:
we might as well be inside a sanctuary or a chapel, which doesn't
mean there's not cacophony here. If the mammal room had begun to
propel me inward, this was just a prelude to the crowning, swooning
achievement of a Hubbardian mood: Hubbard's bird dioramas aren't
just a part of the vast room in which they lodge. They are, themselves, an integral architecture. There's so much in the room with

which they could compete: what feels like trillions of taxidermied birds that have nothing to do with the Hubbard dioramas, including a bird so gargantuan—a stuffed (double-wattled) cassowary—you could be inadvertently sideswiped by its tail (plate 4).

Nothing could be more antithetical to Hubbard's relationship to bird life than the Victorian menagerie of endlessly hatched birds in profile (plate 5), one stuffed specimen per branch inside a case, fastened like tinseled ornaments on an Xmas tree.

Nothing could be less like the mood of this room than the bird that seems nailed to a wall, caught beneath an artificial light toward which it seems to crane, a once beautiful rhythm of life morphed into a genus of ungainliness and kitsch: it still struggles inside a hunter's tooth of steel. Yet, in spite of all the quiet noisy birds that could disrupt this scene and crowd it out, Hubbard's bird dioramas exert a mood, an air, regally secular and serene. Unbeknownst to any guidebook—but it would be unlikely to be mentioned there—the L. C. Bates Museum harbors mood rooms of magical proportions.

Parts of the bird room are lined with vertically disposed niches for northern bobwhite, ruffed grouse, eastern bluebird, and great blue heron, to name a few: all told, in the early 1920s, Hubbard had created twenty-two bird and ten mammal dioramas in the course of three years. The curvilinear background of these hollowed out spaces gives the effect of a reliquary made especially for saints, making the panoply of other specimens in the room seem naked as models in a hobby shop.

In each literal and figurative case, the rhythms of the colors of the landscapes match the tint and textures of the individual bird's feathers, and even though each seriated alcove is built to the same dimensions, Hubbard found a way to balance and alter each scene to suit the bird so that each seems perfectly fitted to its niche in spite of their vastly different sizes.

Another range of birds in a part of the room's middle space—the eagles, the hawks, owls, gulls, and ducks—instead of being vertically disposed, move crosswise and laterally like elephant folios for a book of nature. They have the same hushed graininess of an N. C. Wyeth children's book panel. I'm alone in this library and surrounded by a Hubbard-fashioned evening mood with waterfowl whose diorama poses not just birds but also questions: like, where do birds that live on water sleep? On water or on land? Where do birds live when they aren't present to human consciousness?

In one part of the museum there are definitional lists that are probably not likely to stay with a visitor, as in the "Anatomy of a Leaf," broken down into xylem (the leaf's water pipes); stoma (its mechanism for taking in air); and chloroplasts (the leaf's sugar factories). Leaf margins, one placard explains, can be entire (smooth), lobed (bumpy), or toothed. Some leaves are simple—they're the ones that have a single leaf for each leaf stem called the petiole. A compound leaf has several leaflets. A pinnate leaf has a feather shape. A palmate leaf has a fan shape. Leaves might be as difficult to categorize as clouds, but in the room where Hubbard's birds reside, he knows such calibrated differences by ear: to each bird, its leaf or tree; to each painter, his brush stroke. Feathers move in one direction, and brush strokes move in another in the dioramas; sometimes they move in sync. All of the time, for each and every bird, solo in its habiliment or with several of its kind, Hubbard creates a vibration, a coalescence or a dissonance, that calls forth a mood (fig. 15).

Of course, in a strange sort of way, Hubbard gives his birds a home maybe even more resplendent than the one they lived and died in, but I do not mean for that analogy to hold for the boys and girls who gathered here. If I was crying, it was because I was being rubbed in the direction of a palpable brush stroke: not tickled, but enjoined. I had a feeling that, as with View-masters, the mood of an adult extended its long arm back into a childhood one had evidently experienced and yet never known.

~~~~~~~~~

"the woods after autumn. naked trees. lichens. rocks. beech leaves. winter bouquets. stars. snow. birds." A list penned on a notecard by Charles D. Hubbard of "things he is thankful for."

(15. *reappearing as bright song*)

Around 1910, when Charles Daniel Hubbard was thirty-four
years old, either a sea change or a love that was waiting to happen
was kindled for him—at the very least, he experienced an encounter
that would definitively shape the course of his life's work and application of his artistic energies: apparently, Hubbard and Hinckley—
possibly introduced to one another by Hinckley's brother, Edward S.

Hinckley—met in their native Guilford, Connecticut, at the First Congregational Church where they both were members. In no time at all, Hinckley commissioned Hubbard to design a seal for his school, which by then, was in its twenty-third year of operation. The school logo, in the shape of a medallion, or "roundel," predecessor to the beaver totem that Hubbard would also design, features the sleekly curved figure of a barefoot boy so gracefully poised in profile, so tonally attuned, he seems to have emerged from the flowers, trees, rocks, and books that form the scene in which he is set (fig. 16). His legs are crossed like a woman's and his eyes trained on a book that he props on his knee—literally *Moby-Dick* according to one remembrance. Numerous symbolic items are placed around him like seek-and-find objects awaiting the right discoverer of the scene—a plow, a machinist's hammer, a baseball bat—as bearers

(16. *inward turnings and outward correspondences*)

of meaning to the otherwise abstract words that rim the seal to en-
dorse a clunky Good Will recipe:

RELIGION • HOME • EDUCATION • DISCIPLINE • INDUSTRY
• RECREATION

It's a curious assignment to be asked to create a coat of arms representative of an institution and the mood of its dictum's milieu before
really knowing much about it, but by 1915, Hubbard's attachment
to Reverend George W. Hinckley's mission must have continued to
grow since Hinckley invited Hubbard to the campus to employ his
skills in the very design of the museum he aimed to build, initially as
a home for his own developing collection of specimens and artifacts.
In addition to his occasional fall trips—Hubbard served not only as
artist and designer but as lecturer in astronomy and geology as well
as preacher in the campus's Moody Chapel—Hubbard became Good
Will's summer person, its "rusticator"; the place was his launching
pad for hiking, camping, and outdoor painting whose palette would
become the very walls of the rooms that he and Hinckley created
together.

If there's a ledge, it's a modest one; if it's a path, it's short but
winding. Such sites literally come to be named for Hubbard on the
campus, and these could just as well be paeans to the intimate scale
of everything Hubbard made via Hinckley's authorizing seal. Hinckley's plan wasn't so much "large" as it was grand: to found an institution bred of his own principles and with a paucity of funds. When
he starts with the farm he purchases for his purposes in Fairfield
in 1889, its five acres must seem an expanse beyond imagining—
symbolic fodder for many an American dream—but he and Hubbard
become collaborators in the mutual creation of roomlets, nooks, interiors, and interiorities: of birthing grounds for moods.

To get lost in the mood of Hubbard's bird dioramas is to be made
the beneficiary of the work that resulted from his being invited
into a hamlet. It's almost as though Hubbard is the orphan saved
by Hinckley and vice versa as together they create mood rooms and

hollows for children—those youngsters who are without resources, yes—but also for themselves as children passing as adults. For Hubbard, their friendship facilitated an engagement whose arrangement was one of retreat. As their work together continued to evolve, Hubbard required a house of his own there, so he built himself a small lean-to on the east bank of the Kennebec River dubbed Camp Flapjack in honor of the pancakes he took such pleasure in cooking on an open campfire, "rain or shine." Cozy as a tent, the little house also enjoyed an indoor hearth.

In the same spirit of intimate home making, Hinckley, when it came to constructing houses for his charges, did not rely on more conventional institution-style dormitories. He insisted, instead, on building small cottages for the boys and girls. Nor was the museum the higgledy-piggledy not-without-its-charm ramshackle potpourri that it is today but was conceived with a unique sense of the relationship between intimacy, space, and meditative study.

Hinckley seems to have had an inventive sense of museology, and he wrote about his feeling that the architecture of a natural history museum should not outshine its displays by being inappropriately lavish in scale; nor did he think exhibits should be grouped in such a way to "distract or confuse" visitors as they "rush from one department to another" (GWR, June 1949). Thus, rather than "enlarge" the Bates Museum when, in the 1930s, the material in its exhibits began to outgrow the designated space, Hinckley set Hubbard the task of, in one case, repurposing a building (the woodworking shop) and otherwise designing at least two small one-story wood-framed houses, a little larger than huts, each of which focused on a particular natural history theme: Dendrology (for the exhibit of wood and the study of trees); Granite House (for the study of geology); and Shell House (for conchology).

Good Will, in its design, at least, starts to feel like a form of fairyland headed by gnomes. "Mr. Hubbard is not an architect by profession; he is an artist," Hinckley announced in one of his official reports, the ongoing published record of any and all developments of the home and school. He seems proud of the fact of his audacity

(17. *like stepping directly into a diorama*)

in asking an artist to build a house and to trust what will emerge. For
Shell House, Hubbard carved scallop motifs into humble capitals
that framed the entrance doorway, while Granite House "[was] built
of stone,—the walls three feet thick, at the base, and rising about
four feet above the surface; the remainder of the walls [were] of
logs, after the manner of log cabins" (*GWR*, July 1949). All of these
houses offered a combination of specimens and paintings made expressly for their settings. Stepping into one must have felt like stepping directly into a diorama (fig. 17).

Shell House included four murals by Hubbard of Maine seacoast
scenes, while Dendrology House featured specimens of thirty-three
varieties of trees growing at Good Will, intermingled with four
large-scale paintings by Hubbard unintentionally reminiscent of a
cycle by Vivaldi: "one of an apple tree in spring, one of an elm tree
in the summer, one of an oak tree in the fall, and one of a grove of
pines in the winter" (*GWR*, July 1949). I wouldn't doubt that each
of these houses contained at least one piece of highly personalized
curiosa like the painting Hinckley describes that Hubbard made to

match a childhood feat of Hinckley's of which he apparently continued to be proud. "The painting [made for Dendrology House] was of five elm trees as they appeared on State Street in Guilford more than 60 years after they were planted," and they are part of an ever-deepening environmentalist's plot: evidently, Hinckley won a contest posed to boys of his town by the man who originated Arbor Day and Bird Day (does that even exist anymore?), a man with the name of Dr. Birdsey Grant Northrup (*GWR*, May 1949). The boys were charged with planting five shade trees one spring in 1876, and whoever's were still alive the following September would win a prize of either one dollar in cash or a copy of his book, *Education Abroad*. Hinckley chose the book.

~~~~~~~~~~

Imagine Hubbard carving and gilding his own frames; imagine him and Hinckley "travel[ing] together by automobile for more than three thousand miles" to secure the native New England rock specimens and the sketches for the oil paintings of the slate, limestone, granite, and feldspar quarries that would constitute Granite House (*GWR*, July 1949). Imagine them, too, traveling to New York City, as they no doubt did, separately and together, where art historian Marius B. Péladeau believes they most likely would have witnessed New York City's great unveiling in its newly inaugurated American Museum of Natural History, inside its vastly vaulted and towering halls, the magisterially proportioned and sleekly realistic dioramas that would be the standard-bearers of their day. Fanfare and heyday must have exerted a tide that was hard to resist, but Hubbard's intimate atmospheres, by turns, enfolding and opening like the feathers they embodied, couldn't be further, aesthetically and conceptually, from the larger field that set the terms for habitat construction.

Hubbard isn't a Carl Akeley or Teddy Roosevelt strangling leopards in the wild with bare hands or felling gorillas and elephants in far-off lands. Entirely unheroic, his work isn't at all interested in the trophy, the kill, or the hunt. It is born of the nowness of the near-to-hand, from a vantage point that is the opposite of declamatory and just the size and shape of a person.

The L. C. Bates Museum is tuned to a local localness, and its displays are like a quilt the entire community is bent over, made from the sights and sounds, the hides and haunts of the animals that are its fellow inhabitants. "Without a doubt it spent part of its life on Good Will," Hinckley said of a wild cat found two miles from the museum by one Gus Gustafson who killed the creature with a "cant hook" when he found it attacking his dog (*GWR*, March 1949). All but three of the localities of the dioramas are set in Maine, with several set expressly on Good Will's land, and most of the specimens, like Gus Gustafson's, were gifts: in 1931, a Good Will graduate, Mrs. Marjorie McGuire donated one of the last caribou shot in Maine; a Mr. Evans Cobb donated a calico deer; people brought birds that had been found dead or had been killed, and Hubbard traveled to the spot to experience the surround before he painted it. Not exactly folksy, the museum could probably be considered a family affair, but the thing that kept it from being claustrophobic, cloying, or parochial was Hubbard's dedication to *air*.

If I were to try to encase Hubbard's mood making into a nutshell made of seemingly opposing sides, I'd say that he was a plein air painter who made enclosures. His preference for working in the open air is crucial, not only for the fact of the hundreds of resulting small thumb-box oil sketches that Péladeau judges to be just as fine as any of his large-scale works of art, but because of its enabling him to work with his ears as much as with his eyes. Rather than picturing or staging birds at flight or birds at rest with the effect of freezing them, he steeps his subjects in rapidly or slowly moving chords of color, environing each bird by way of its call, effecting a synesthesia of sound and light.

A defining method of conventional habitat dioramas, which, without a doubt, requires a great deal of skill to perform, is a kind of geometric trick that exploits an optical illusion: the back wall of such structures is curved, but it is painted upon in such a way as to make it seem like a flat expanse. In Hubbard's bird dioramas, the curvature of the presumed "background" is visible since his method was to enlarge his thumb-box oil sketches onto a flexible canvas that could then be rounded into a semicircular concavity inside the case (*CDH*, 9). Rather than trick us in the service of conventional trompe l'oeil, the effect is to invite us to enter an atmosphere that mimics the bowl of a bell, oriented neither by foreground nor by background but mood sound.

It's not surprising to learn that Hubbard would recite poetry— Dickinson, Longfellow, Frost—to his students while they silently drew (*CDH*, 11), or to discover him listening for the sound of words with the same intensity that he listens to the sounds of animals, trees, and birds. A book like his 1952, 109-page *Camping in the New England Mountains*, "dedicated to Doctor George Walter Hinckley through whose friendship [he] was enabled to behold the mountains that are round about New England," is remarkable for the dedicated fluency of his gorgeous hand lettering, perfectly right justified and intuitively timed to match the tempo of composition with the lengths and heights of words themselves; for the combination of finely indented illustrations in pen and ink, square or oblong, squat or slender, inlaid inside the paragraphs; for still other full-page drawings not meant to illustrate an accompanying narrative but offered expressly as "creative works intended to give the *atmosphere* of Camping in the New England Mountains"; and for their author's attention to sound (fig.18).

The sound of the word "rucksack" itself is soothing, especially if it is inscribed with the insignia of Hubbard's heart-shaped initials—*C* arcing to enclose *D* set beside a bosomy *H*; especially if it is rumpled like the blanket on which he lies while he listens to "the bedtime music of the stream." Intermingled with a recipe for porcupine and cranberry sauce prepared camp-side, Hubbard listens for "frogs, whippoorwills, and perhaps the thrush." Climbing and pausing to watch of an afternoon, he finds a world "all ahum with music of bees." Climbing higher toward nightfall, he hears hints of the crepuscular in "the bark of the fox," or "the scream of the panther," "the hush" of a "brooding silence." Not "the cribbing of a horse," that

at Camp Flapjack was spent in sleeping and eating. I had come
to Maine in the first place for other reasons. Doctor George W.
Hinckley of the Good Will Home Association had engaged me
to paint landscape backgrounds for the habitat groups in the
Natural History Museum there and so it came to pass that
there were trips taken to Mount Desert Island after studies for
 the sea bird colony,
trips to Mount Kineo,
trips to the quarries
of Maine for specimens
of granite, of limestone,
of slate, and trips to the
sources of the various
minerals after material
for a geological display,

always returning to Camp
Flapjack, my home camp.
Then, too, there were my own
personal journeys for sketch-
ing to Ktaadn, to Boarstone,
to Tumbledown, to Bigelow,

and to various localities all along Maine's wonderful coast.
Among my neighbors on the east bank of the Kennebec
was a herd of Ayrshires and, as cattle have always been interesting subjects to me, I frequently rowed down stream to the more

79

(18. *the campfire's mood, the angler's ambiance*)

sound is of a rippling streamlet; lost in blue haze, his "pulse quick-
ening in direct proportion to the proximity of contour lines," he's
tuned to hear a "viveo" that "sings somewhere in [this] sanctuary."
Bald ledges sport bald eagles, their girth and piqued brow cut by the
scent of "fly ointment."

The word "swale" puts me in mind of things that swell and sway, whereas "banded gneiss" with its oddly silent *g* is nothing like the hardness of the layered rock it points to. I'm as mesmerized in his prose by the sound of the word "hornblende" as I'm drawn to imagine the fact of its color—deep black; of the word "tourmaline" and its hue of green-set-in-garnet. On one part of a trek, to hear the "Ah ce ce ce of a white-throated sparrow or the honking of wild geese" is "enough," he says. At another point along a different pass, he is "bewitched simply with the names scattered over [his] maps." An entire history of native inhabitant and European conquest floats inside the seemingly placid drumroll of two words: "birch and bateau." Showers drum upon the surface of a tent beneath a night sky's "interstellar rambles." The voices of strangers singing inside this "vast theater" drum upon the surface of his text, reaching him, not far enough away.

I hear or read that owls can stand on a limb of a tree and pick out a mouse's heartbeat under the snow. Once, through my kitchen window in Providence, Rhode Island, I gasped to see a barred owl in a nearby tree. The bird was so large and so unexpected, at first I thought it was a cat. There's probably no such thing as a diorama that doesn't dwarf the wild creatures it contains, implicitly taming them no matter the dioramist's attempts to simulate ferocity or stealth, predatory power and maw. Pulseless. Soundless. These are the bare facts of natural history displays, but I'd swear in Hubbard's rooms I hear, not the heartbeat of a mouse, but of an owl, uncertain if I'm confusing it with my own. Up close, the feathers of a snowy owl are like bearded fur, and I take the plunge and lose my footing to autumnal streaks of yellow and shrieks of golden brown (plate 8).

~~~~~~~~~~

Painted backgrounds as accompanist to taxonomic specimens go as far back as the late eighteenth century with Charles Wilson Peale, but the habitat diorama, gaining its first foothold in particular in Sweden and the United States, is understood to be a late nineteenth-century phenomenon. Early, preeminent ornithologist, Frank M.

Chapman (1864–1945) was the first in that era to situate birds against scientifically accurate painted backgrounds, and it was he who initiated the practice of trekking to the site where the bird lived and died.

As the curator and designer of the paradigmatic Hall of North American Birds in 1902, he directed a team of "artists to document, record, and collect specimens, accessories, and landscape scenes," with his choices often based on species of birds that were "most imperiled."

An exact contemporary of Hubbard, Chapman offered a step-by-step, developmentally staged imagining of what should happen for a visitor to the halls where his work famously appeared—in the American Museum of Natural History in New York—that couldn't be further from Hubbard's beautifully eccentric creations and their gestalt: "The aimless visitor involuntarily pauses. His imagination is stirred, his interest aroused, and the way is opened for him to receive the facts the exhibit is designed to convey."

Either an adjunct to a scientific society *or* a means of popularizing science; *either* an instructional display *or* a pleasure palace; *either* an accurate reflection of nature *or* a subjective interpretation of the world; *either* a sense of the presence of the artist *or* foregrounding of the facts; *either* a Vermeer-like devotion *or* a brief van Gogh passion. The mutual exclusivity of science and art, their clash and cancel, their unquestioned rivalry: Hubbard is an outlier to the weary bifurcations that seem to have attended the history of the habitat diorama in its most heavily institutional contexts and preserves.

Given the peculiarly out-back, off-the-map homey nature of the L. C. Bates Museum and the larger project of which it was a part, it wasn't likely for Hubbard's work to fall within the purview of Karen Wonders's invaluable studies of the habitat diorama, thus leading her, understandably, to assert that "none of the early American diorama painters had an artistic training whereby they could apply the latest knowledge of Impressionism to their particular genre. Even had they tried to do so in the scientific context of the natural history museum, it might have presented a problem." I can create a kinship between Hubbard with German-born wildlife artist Carl Ringius, if not in style at least in spirit, given the controversy that attended his 1939 background for the famous fighting Alaskan moose, commissioned for their installation in the Hall of North American Mammals in the American Museum of Natural History. Whether his brush strokes were considered "loose," "impressionistic," or "subjective," the attendant assumption—which I find quite wonderful—was that Carl Ringius's painting remained unfinished.

Finish is also something that Hubbard's mood creations lack, though they are clearly and consummately complete. Hubbard obviously doesn't subscribe to requisite "quiet tints that don't attract away from the specimens" in part because his pieces are not about the birds *as* specimens. If I experience the mood of a heartbeat in the one thousand four hundred and eighty-five square feet allocated to the birds, it's because of his invitation to collocate bird and human as part of something greater than themselves. The staccato brush stroke speaks to a reverb of beats in the form of sculpted correspondences that are part of each of the diorama's charms: the woodpecker and its pith (plate 6); the bird *as* the pulp of the tree that it is foraging; the pattern of algae on a tree stump and the sum register of blues and greens in its bird, the eastern blue bird; the z-shaped movement of a green-backed heron and the indented switchback of a shore; the graceful footing of the ruddy turnstone and the moving shore upon which other birds are painted-in; the plump white toadstool protruding from a birch tree and the squat-ly shelter-colored spruce grouse.

Not a "window onto nature," these are windows onto imaginary rhythms; not intent on making us feel that we are "standing before an actual scene," these enveloping views hint us hitherward toward the actuality of a feeling. Mightn't it be possible for a habitat diorama to elucidate an interrelationship among organisms at the same time that it activates a relationship between conscious and unconscious states? Hubbard wasn't trying to create illusions capable of transporting viewers to (actual) places where they had never been as a sort of armchair tour guide. Like Florence Thomas's creations, his interiors take us to places that bear traces of a highly privatized world we know but don't have access to.

~~~~~~~~~~~

The bird identified as a Merlin in Hubbard's series offers an interesting case in point. The "background" painting in this instance includes painted-in shadows of the specimen and the branches on its perch. Such shadows are impossible given the bird's size and altitude and distance from the mountains; they are unnatural; they are also painterly. They leave us to question the bird's location in space and time and the landscape's too. They make us wonder where the light source would have to be that could cast such a shadow. They tell us "you are here," but where? Not interested in accuracy, the Merlin diorama points up the otherworldliness of the world. The panes of the windows of other dioramas are segmented, thus interrupting a seamless or illusionistic view. Set inside such dots and dashes, a bird or a branch is a shape in the landscape, and the diorama is as much about birds as it is about light.

Though Frank M. Chapman, Hubbard's more famously esteemed counterpart in New York, was personally unknown to Hubbard, the two meet at an intimately opposing crossroads in this history: Chapman places shorebirds and bald eagles and so does Hubbard. A rare moment of a shorebird in flight, stage right, might almost mimic a similar arrangement by Chapman, but there the similarity ends. All of the assemblages in Chapman's Hall of North American Birds are squat aquariums that reduce the power of a bird of prey to a human waistline at eye level. We should never be inclined to look down at an eagle in midflight so proximate to our body we could touch it. Another bald eagle apprehended from behind seems tame as a heart-shaped box of Whitman chocolates; familiar as the city pigeons and as liable to be chased by reckless children; as obediently available as a gerbil in its basement cage (fig. 19).

Neat, stark, pristine, the painted backgrounds in Chapman's displays use color as sheen, the better to reflect the birds as waxen outlines, whereas Hubbard's birds are sunk, enmeshed, embedded in absorptive planes. If I search out an analogy for a quality of light, I can't get past the idea of the background paintings of the better-known shorebird displays resembling a 1970s Chevy fender even though their horizons were manufactured decades earlier than that. An invented kitsch-o-rama or a scene suitable for a thousand-piece puzzle, these reproductions are plastic and metallic in their feeling whereas Hubbard uses color like water, air, and wood. Even if Hubbard's eagle could be mistaken at first glance for a kind of anthropomorphized profile in courage or mascot of a nation state, the bird is one equivalency inside a landscape that rises while an ocean recedes; in which an ocean doubles as sky while a rock face tumbles downward (fig. 20).

Does Chapman's Hall, comparatively, create a mood of death? The ducks and gulls are as realistic as skeletons. Bands of rock birthed by sea spray are the home for herring, gull and common tern, for hooded merganser and blue-winged teal: Hubbard's rooms create a mood of life (figs. 21, 22).

Hush. It's snowing in there, and out here too, in the woods that create an enclosure for the northern goshawk which might just be my favorite of Hubbard's work (fig. 23). This must be one for which he painted his cabinet's light source blue to create a sense of nightfall. The sound of a "puh-puh," of snow sliding from branches that are warming now onto the ground below, the sound of "pit-pit" of snow falling into itself chilled and startling, brushes me into the space of a quiet more original to my being than I am to night or day. Where do birds go when it snows? They sleep inside the whiteness or keep their eyes open to its sheltering silence. Both distinct from and part of their surroundings, capable of disappearing into it, then reappearing as bright song; absent, present, light and dense, immanent: birds are like this, as are moods. And life.

~~~~~~~~~

When I tell a friend who is an intellectual about the funny admixture of a Hinckley-Hubbard concoction, he summarizes their

(19. *as obediently available as a gerbil in its basement cage*)

relationship perfectly: "So you're saying one was like Dickens, and the other like Proust." "That's it!" I say. "That's exactly it!" On one hand, we have a Dickens, bent over a social problem novel of Oliver Twists, and on the other hand, we have Proust, dedicated to the mercurial wendings of childhood weathers and adult memory maps, haunted by whiffs impossible to capture. It's hard to imagine the two sharing the same room let alone building a project together.

I don't know if George W. Hinckley exactly lorded over the premises of his homes, school, and museum, but the tone of his personality lurks around every corner in a way that is hard to warm to. It's probably impossible for an orphanage built up out of one's own individual sense of righteousness not to emit an air of eerie cultishness, and aren't such projects often enough as much about the benefactor as those whom he wishes to help? Hinckley's dictums and seals, his turgid texts are all a bit too earnest, too self-consciously upstanding; they suffer an off-putting combination of the simpering and the authoritative, none of which suggests he didn't change children's lives.

(20. sunk, enmeshed, embedded in absorptive planes)

It's just that the slant of Hinckley's oddness isn't a hill I'm compelled to climb even though his ties with Hubbard, or Hubbard's ability to work with it, tinted the elixir that created the captivating mood rooms "of which I write."

In 1943, having arrived at the age of ninety, Hinckley composed a book-length account of his life, eccentrically titled, *The Man of Whom I Write*. In an apparent attempt to surmount his egotism, he says, he decided to replace each and every instance of the pronoun "I" in the book with this clunkily inverted six-word phrase. Needless to say, the effect of encountering "The Man of Whom I Write" at every turn makes the book more, not less, self-indulgent, to say nothing of its effect on the prose's reader-friendliness. It's not just his curiously inexplicable German accent and bushy moustache or his closely cribbed wire-rimmed spectacles that make it hard for me to imagine him singing around the campfire with the unloved children in his care. It's his inability to tell a story (here the analogy with Dickens fails to hold): his utter lack of timing; his slogging

(21. *neat, stark, pristine*)

rigidity, and tendency toward stiffly aphoristic platitudes that makes
me imagine his kids more often snoring than standing at loving attention in his midst.

In the 1860s, Hinckley was himself just in his teens when he met a
boy his age newly enrolled at the Guilford Institute in Connecticut—
a boy sent there by the New Haven Orphan Asylum named Ben Mason. Not only was Ben noticeable on campus for being new to the
school but, to Hinckley's mind, for the physical figure he struck,
which he describes in detail: "He was slender, boyish, aggressive; he
had red hair, blue eyes, freckles; he wore a butternut-brown jacket
with brass buttons, a cherry-red necktie, navy blue short pants,
and black stockings." It's not clear from the text whether it's Ben's
looks and attire, or his situation in life—he was both fatherless
and motherless—that most drew Hinckley, but the prose becomes

(22. *what makes us quake or makes us sing*)

breathless just at the point of this gathering of details as he designates Ben Mason, in a bound, as "the most interesting person [he] had ever met."

Hinckley determines to seek Ben Mason's friendship, and before long, Ben, confiding his heartache, stirs Hinckley's sympathies until arriving at the "thrill" of a particular "moment" that Hinckley says he relives perpetually: that instance "on the corner of York Street and State Street" when he offers Ben Mason to come live with him in his father's farmhouse even if it requires that Hinckley leave home. Though he offers Ben this prospect before asking his parents, they agree to take Ben in, inaugurating a period during which Ben and George are "inseparable night and day." They worked together; attended church together, "roamed the woods together; visiting the swimming hole often; read poetry to each other in odd moments."

(23. *Hush. It's snowing in there.*)

The house really isn't large enough, and eventually Hinckley does determine to go, that "Ben should stay in [his] place."

Ben Mason comes to signify the "class of boys to whose welfare [Hinckley] was to devote [his] life." When Ben Mason passes the New York bar exam and becomes a successful lawyer, his story becomes ever more useful; when Hinckley suffers periods of longing, wondering how Ben is and whether they will resume long-lost correspondence, he recounts a journey on a train through the very area where the swimming hole lay. Hoping to get a glimpse of his and Ben's old stomping ground through the train window, he is greeted instead by the daily papers announcing Ben's untimely death at the age of forty-one. The 1901 news story is titled "Died a Victor" and tells of a lawyer who, soon after calling his wife from out of town to tell her that he'd won a case, returned to his hotel room and promptly died.

A painting of Ben could, I suppose, function in the way the Good Will's branding roundels did—as a sort of advertising seal of the

(24. *"the most interesting person he'd ever met"*)

home's mission (figs. 24 and 25). Not to have figured Ben Mason as the lawyer he became seems to arrest him in that moment of pre-adolescence that, for lack of a better understanding, kindles the fires of something verging on "the cult of the boy," as much a turn-of-the-century movement as a privately invested Hinckley/Hubbard affair.

Drawn by Charles D. Hubbard.

(25. *in a cupboard marked "the mystery of mood"*)

Did Hubbard rely on something like a school photo in creating the portrait in oils of Ben that hangs in the museum, decades after the fact of George and Ben's meeting? Or did the painting's fieldwork depend on conversations with Hinckley? "Describe him to me," I imagine Hubbard to have implored. "Describe to me the love of your boyhood days." Though the cherry-red necktie does not appear in Hubbard's painting, the brown jacket does, and it is worn by a fiery-eyed boy whose cheeks and lips are ruby red, whose thick dollops of hair are reddish-brown, and who emerges against a background of indeterminate reds and yellows.

I can't know what Hinckley may have thought he saw when he described Ben Mason, who was, in fact, a boy, as boy*ish* in appearance. Is the fervor that drives the project aimed to meet the needs of a particular boy, starting with Ben Mason, or is it set to the dial of a harder-to-define, once felt but forever out-of-view mood of being cared for in a way no home or school could yet provide?

Who knew that a museum could be a place for conjuring a hazy realm of sensuous nonknowledge. Or that houses for taxidermied creatures could supply a mood of what, to their creators' minds, it could feel like to be loved.

~~~~~~~~

"The ten commandments and a handicraft make a good and wholesome equipment to commence life with" (*GWR*, May 1949). That's Hinckley quoting Dr. Birdsey Grant Northrup quoting nineteenth-century British historian James Anthony Froude. Imagine replacing the word "handicraft" with "taxidermy" in that motto, and then imagine the taxidermy in Hinckley's museum being created by one of Hinckley's orphan boys. Imagine the forms that instruction would take: in the morning, "Honor thy father and thy mother," but you have no father or mother, or if you did, you were abandoned or neglected by them; in the evening, "Do not haggle the skin at its margin under the antler burrs." Imagine an actual child in Hinckley's care, and one who wasn't just the mood rooms' intended audience or recipient, but in some sense one of its cocreators.

A figurative descendent of Ben Mason, Walter Reeves (fig. 26), a graduate of the Good Will system, even better, "one of the first boys at Hinckley," came during his time there, in the late nineteenth century, no doubt when he was just a boy, to learn the art and craft of taxidermy from Mr. W. R. Gifford of nearby Skowhegan, Maine. In his day, Gifford's business card offered cash for "birds, animals, game heads, horns, fur skins, freaks of nature, and living animals." Reeves is depicted as "the first boy to make a great response to Hinckley's small museum." This would have been around 1890 when the "museum" was not more than a room in Prospect Cottage. By 1896, the room expands and the collection is moved into Moody School, with access through the library. Hinckley describes a mounted loon

(26. *the room all misty, the floorboards scattered with excelsior and dried blood*)

that resided there, "captured and mounted by Reeves . . . and indeed other excellent specimens of his work are seen on every hand." Would Reeves had been encouraged to kill a loon for this purpose, or are we meant to understand by the word "capture" Walter Reeve's expertise in conveying the spirit, the nature, or mood of that beautiful bird? When this "museum" was devastated by fire in 1904, Hinckley was only spurred on to envision a museum building all to itself—the entity that, by 1915 and on into the early 1920s, would emerge as the L. C. Bates Museum to which Walter Reeves is also said to have contributed: his teacher, after all, Mr. Gifford, had mounted the infamous caribou.

"Labor develops inventive talent" (*GWR*, May 1949) rings as a motto for any Good Will boy, so we have to try to imagine Walter being offered this practicum in stretching, relaxing, stuffing, or mounting dead skins as a building of character. I imagine the room all misty, the floorboards scattered with excelsior and dried blood, when Gifford instructs him "to lay the animal on its back," "unjoint the bones," "skin out sections," or "to take all the flesh out and take out the brain." Is Walter meant to become like Spinoza, who, again, cited by Hinckley cited by Northrup cited by Froude, conjoined the power of intellect with the acquisition of a practical skill, choosing to earn a living grinding object glasses for microscopes and telescopes rather than accept pensions that were offered him (*GWH*, May 1949)? I know nothing of the intellectual prowess that Walter Reeves may have developed at Good Will, but it appears that taxidermy became his life, and a no doubt lucrative life at that, since he and a brother maintained a taxidermy practice in Portland that even included "novelties in taxidermy" among their suite of offerings, and he was described as being connected to the largest taxidermy firm in New York.

What could it have been like for Walter Reeves to have arrived in a state of lostness and fear that parentlessness can signify and to have gotten the gift of taxidermy out of the deal? If he were only creating specimens denuded of their contexts, that would be one thing, and maybe this is as far as it went, but, insofar as his growing up occurred in tandem with the growth and development of Hinckley's "museums," serving even as a boy-contributor to their displays, I like to imagine him as participant with Hubbard in the creation of the habitat dioramas. To tell the truth, we might, in the end have to dispense with the category altogether, for what if Hubbard wasn't really making habitat dioramas after all? What if he was just being inventive and resourceful, his inventiveness an outgrowth of a dedicated labor? Hubbard used *the idea* of the natural history museum to make *paintings* into which three-dimensional objects could be inserted. He made paintings out of cabinets and goose down. Walter Reeves's taxidermy, were it to contribute to such scenes, would have to represent something more and different than the reformed husk of what once was a living creature. It would be one element in a painting quavering with dimensions advancing and receding of living and losing, of boyhood moods, and adult recapturings.

Even as I write these sentences, I'm chilled to the bone by the sound of a furred fluttering inside the walls. It only happens after evening fall in the cabin, just adjacent to the desk at which I scribble and scratch. I know it's a mammal, not a bird, that busies itself in the lip of the enclosure that is my house, that warming space between an outer and inner carapace, but its tumbledown thrum recommends folded wings, or practice flights in airtight atmospheres. At first there's this sound of soft and steady flapping along a hidden gauntlet, then a thump and skitter. Animals skulk, slither, and dart. They bristle. Only in cartoons made by adults for children do they prance and hop; only in fairy tales do they bounce rather than pounce. Is it possible to ever know them if it's really our own humanness we fear?

Tapping a taxonomy of Hinckley kids, any one of us could find ourselves there: no doubt there were those who were not wanted, and those who were improperly cared for; those whose parents had died, and those whose parents had simply disappeared; those who were sole survivors, and those who began without a past, left on a doorstep or pushed into a street; those who remembered their parents, and those who forgot; those who afflicted a burden beyond their parents' means and those who were their parents' means; those who forgot how to eat, and those who forgot how to sleep. Were there many there like me?—the children of violent men?

~~~~~~~~~~

I think of our developing moods as made of a fabric, tissue thin, incompatible with violent acts, and certainly not born of them. A diaphony that can suffer a rip, scratch, or tear, and then what? Or is mood a baseline hum immune to circumstance that can't be tampered with: a disposition? At any rate, where children are concerned, and maybe grown-ups too, people are scarier than animals, and to some extent we are all hidebound and bred by our fellow humans' unpredictable moods and their wild ways, no offense to animals.

Bats don't really have any interest in nesting in the hair on your head, an eagle isn't likely to crack the skull of a person, spiders don't bite people when they are sleeping—they really want nothing to do with people and prefer a warm dark place for spinning in. It's not all about you. Fellow humans, though, can and do get under each other's skin, unshouldering us at a baseline, rendering each other suddenly spineless, taking each other's breath away for better or worse. Neighborhood dogs aren't the problem; people are. Dogs can be reasoned with, a dog can be reached, but when my father was in one of his rages, he was simply all tooth, fang, and nail, he was gnarl, and things breaking around you, and the thought that if you didn't cower, he could strike or bite you, so you spun yourself into a frozen form: taxidermied.

In a sad turn of fate, my father spent the last decade of his life wrapped in the tourniquet of Parkinson's disease, stuck in the space between tremor and paralysis, twisted into spasm. There really are no words for the quaking and the helplessness bred by the disease, and what they don't tell you is that, when the medications stop working, the symptoms they were meant to calm return a thousandfold. I was convinced by the state I'd find my father in that the inhuman ravages of the disease couldn't be what everyone with Parkinson's endured but only people in the working class whose doctors, I imagined, as neglectful and uncaring, whose word my father implicitly trusted, who confused "quality" in "quality of life" with buying power: quality was something you need qualify for.

The nursing home where he spent his last days sat as though it were awaiting him all these years though he barely would have noticed it in good health. The "old folks home," as we called it when we were kids, lay at the bottom of the street a short two blocks down from the hospital where I was born. As a teenager, I volunteered there for a spell—I played music for the "inmates"; I joined them in checkers and cards; mostly, I listened when they sought an audience for the stories of their lives. Decades later, it was as though I were handed a strap and slapped into the fact of the brutality of age. Yes, it was as if I'd been given a strap and instructions to go in and tie my uncontrollable father down, with the additional warning, "Don't forget to wear gloves."

Even though "mood" and the multidimensional matrix that goes by the name of "self-control" are linked, I never thought of my father in his younger years as suffering from a mood disorder. The glum consistency of his demeanor actually seemed in keeping with his unpredictable outbursts, which, however passionate, were equally dark. Perhaps, in the 1920s of his childhood, there was no such thing as time-outs for tantrums; there were no mood rooms in his house, so he never gained any measure of self-soothing. Now, in the hands of Parkinson's, he was confined to a bed turned storm-tossed sea. At his worst, the bed could hardly hold him.

I might find him curled into a ball of medically induced hallucinations that had him pawing at imaginary cobwebs or repeatedly pressing with his index finger into his eyetooth as though there were something other than a part of himself extruding from his mouth. It was as though he were trying and failing to find a one-to-one correspondence between one part of his body and another. Or I might find him so wracked with tremors that the bars on his bed were raised on all four sides to keep him from falling out of it.

On one such day, he insisted on sitting up—no matter how weakened he'd become, he still was capable of yelling—and I risked it, even though every part of his body including his head was flailing, I released the side bar of the bed and propped him upright. I braced his legs between my own; I positioned my head alongside his head to straighten it; I wrapped his torso in my arms. Of course I was stunned that day into a recognition of his size: not that the man who had once lorded over us, cracking the ground as well as my brothers' flesh with the whip of his rage, was dependent now upon my strength, my leverage, my command; not that he was now small enough to be held in place, but that he had always been small, even petite. It's only the child mind that makes parents into giants. My father had always been small; who knows if in his smallness, there had been touches all his lifetime of boyish beauty.

~~~~~~~~~~

That day, I'd also brought a gift, though my father was incapable of receiving gifts. Nothing you could gift him was something he wanted or needed or enjoyed, though I do recall hitting the jackpot circa 1972 when I gave him a T-shirt with a picture of a cigar-wielding Carroll O'Connor on it, and the slogan, "Archie Bunker for President." In the last twenty years of his life, my father had started to write short stories as well as nonfiction reminiscences. He seemed driven by the sudden need to produce narrative upon narrative upon narrative. I determined that the thing he'd most enjoy would be the gift of one of his own stories, so, as I held him there, I read from the pile of pages that Jean held steadily before my gaze. It seemed obvious to me and right that the thing he might most appreciate and want was the sound of his own voice coming back to him, the tender fact of the matter of a reader. His frozen jaw relaxed and he lifted his lips into a smile. He even laughed—at his own jokes; he whispered, "thank you," and said he hoped I liked the story too.

During the same period of making stories, my father took to using the computer to generate lists—like personal catalogs, they were collections of photographs oriented around a single theme of his choosing. Though he'd always been perhaps mostly of necessity a penny pincher, he thought nothing of using his color printer to support this new habit. It was sad to think that this is all he could come up with to squander his retirement savings on after forty-five years of hard labor at the Philadelphia Naval Shipyard. He didn't just make one forty-page printout with thirteen or so color thumbnail images per page, but at least a dozen, since he also sent them to people in the mail—to me and my brothers, to my mother from whom he was divorced, to his girlfriend, to his sisters and brothers.

Sometimes the sets of images also included brief captions of his own devising; sometimes he also enclosed collections of types of phrases and sentences, for example, ten pages of his favorite tongue twisters, like: "Of all the felt I ever felt, I never felt a piece of felt, as that felt felt which felt as fine, when first I felt that felt hat's felt." Or a list of proverbs, including his own favorite, "If it ain't broke, don't fix it." (Is that really a proverb, I want to ask?) And his father's favorite Sicilian proverb, "The pig always dreams of the squalor it lives in." Not only do I fail to understand this proverb, but Google can't discover it in its memory banks. Usually, he would enclose a brief letter as follows:

Dear Mary and Jeannie:

How is everything going? Some of my teeth feel loose with a minimal amount of pain and discomfort. My guess is more yanks.

Enclosed are a collection of pictures with comments on some of them. Needless to say, pictures tell their own stories. Some of them may be pleasing to the eye. The world is loaded with everything from the very grotesque to the most stunningly beautiful.

Love, Dad

This particular packet includes a collection of "Bugs" (mostly beetles with colorful shells and butterflies), "Hummers" (as in hummingbirds, also quite beautiful), and "Grotesque Sea Creatures," among them the "Hand Fish" (it resembles an amputated human hand scrabbling along the ocean floor); the "Blob Fish" (imagine the face of Uncle Fester smothered in cold cream); the "Fang Toothed" (think of a mouth propped open by jagged sticks for teeth inside a skull backlit in yellow); the "Cabbage Fish" (a gargoyle face turned inside out); and the "Red-Lipped Bat Fish" (it looks the way it sounds).

It would have been better if he'd arranged the pages of pictures to start out with the grotesque and end with the beautiful as his letter suggested, but he hadn't. The declension from beautiful to grotesque instead recapitulated the shape of his moods, unable to linger long on grace or delight before the hideous crept in. In every case, my father asserts himself as the "author" of these collections, heading pages of "optical illusions," or "origami," as "Assembled by Joseph S. Cappello."

In the Facebook age, the impetus to reproduce, collect, and distribute and call that authorship is commonplace, but in the days when I'd receive my father's version of a natural history museum in an envelope through the mail, I experienced them as a strange and inscrutable burden. The man barely ever talked to me, but now I was being made the recipient of his bizarre creations. I thought there must be something to them, though I never went so far as pretending they might in some way be meant for me. That I've saved them rather than line the cat box with them must be a sign of my believing they hinted at some part of a man that never showed itself. Mustn't there be something there worth keeping as a trace of him? Not as a sign of the father whom I wished I'd had—my saving wasn't so sentimental as that. This was evidence of a person all his own, who had nothing to do with me, out of whose mood space my own sense of being in the world had been formed. At the very least, it was something he had made, and somewhere along the line I must have learned that when someone gives you something they made, no matter how unlikely, inexplicable, incongruous, useless, or downright ugly it might be, you had to keep it.

By now it should be clear that our moods depend on what we were encouraged to make or discouraged from making as children as well as on what we allow or disallow ourselves to make as adults. Because my father rarely made things for us, I came to overvalue the things he did make. To my recollection, there were two things he made: one was a facsimile manger to be displayed at Christmas time, a sort of View-Master meets habitat diorama display. He made an enclosure shaped like a barn for the holy family and three wise men. Best of all, he carved out a segment to perfectly hold a lighted star. Better yet, he wired a small light into the inside roof that lit the scene at night. My mother would take it out of its box every year like a priceless chandelier. Into this seasonal diorama, I would pose and stow plastic figurines of animals that sometimes got lost in the paper hay. Their scale was much smaller than the plaster-of-Paris statues, and I seemed intent on making the manger into Noah's ark.

One other time, my father made a teepee for a Native American doll purchased for me, I'm not sure why, at a Western-themed restaurant where we ate once every five years. More boyish than girlish, I was never one to play with dolls, and at the time of the gifting I was too old for dolls. Nor was this doll even the sort that was meant to be played with: it had a stand to which it was secured with an elastic band, as though awaiting its insertion into a collector's display case. The doll inspired my father, and something about the way he went at this project, using a remnant of drafting board from his workplace, applying rulers, and sliding his pencil behind his ear encouraged me to regress. For several weeks, I ceased to be any age at all as I lay on my belly in the living room determined to create scenarios for the lone doll and her teepee, all the while bathed in the sound of the TV my father watched nearby. Though neither of us could say so, the short spell—one week, two?—during which we found ourselves inside the teepee/TV net was the closest my father and I ever came to playing together. I think what brought the prospect of developing the sadly vacuous game to a close was my asking my father if we couldn't make other things—the doll needed peripherals in order for me to continue to "play" with her, for example, a campfire, a loom. When I asked him if we could make these things, he roundly replied, "No." The teepee-doll-TV scenario apparently took on the form of a "case closed." But I never forgot that we had made something worth trying to make again, if not together, then alone: it was shaped like a hollow; it was without words; it was presence without awareness of presence; it was a mutually entrusted mood.

<center>~~~~~~~~~~</center>

No matter how hard we paddle, the wind carries us back; better to drift so as to effect a difference in return.

The day after my first visit to the L. C. Bates Museum with my friend Caren, she wrote with news of another phenomenon she planned to grace me with: "I spoke yesterday about a woman who lived in the 18th C. who took up paper cutting to create 'flower mosaicks' very late in her life. Her name is Mary Granville Pendarves Delany (1700–1788). Gabrielle expects to bring the book this evening to show it to you." I wrote back to say "thanks for a wonderful day, that I do hope will find its way into a mood diorama made of words sometime in the not too distant future . . ." That was in 2011.

One morning in the weeks following our trip, and in spite of Caren's apparent stamina for a twenty-mile bike ride we'd enjoyed the month before, she awoke to a diagnosis of metastatic ovarian cancer. Within a year and half of submission to grueling treatments, she had died, but not before giving her community a literal bridge that she built as an aid to people roaming the marshy footpaths in the woods that ran alongside her house. Not exactly in turn, but in due course, and as she wished it, a group of her lesbian friends built her coffin, and then convened a painting party following Caren's specifications in colors reminiscent of her native Florida (did I mention that plastic pink flamingos sprouted in her Downeast yard?). These same women carried her body in the coffin they had made to her local grave.

After Caren's death, I traveled back to the L. C. Bates Museum alone. If I went in through the backdoor, could I eavesdrop on the memory of my visit with my friend? Could I pick up the thread where our journey had left off? Could I know the Hubbard mood rooms in sickness and in health, for richer or poorer, till death do us part?

Not letting myself be led, I hovered on the outskirts of a tour that Director Deborah Staber was giving to a group of local flip-flopped and T-shirted volunteers who were discovering the museum and helping her sift through some of the material in its annals. "There's a lot of this and that," I heard Deborah say as she moved the group with quiet exuberance from one pocket of surprise to another until each overheard enumeration made me aware of what on my previous visit I had overlooked: here's a fossil of one of the first vascular plants found in Maine, proof perfect of continental drift. Among the oldest items, a most beautiful bird, a passenger pigeon now tragically extinct. The elephant stump that had wracked us with laughter I now understood as a teaching tool: part of a dismantled elephant discarded by Harvard, the Bates displays the limb and exposed excelsior to give children a glimpse into how some taxidermy is made. How had I missed the anomalous antler collection? Especially the deer with a third antler emerging from the corner of its eye? Deborah posed an alarming prospect here: we're not sure how exactly the eye antler would be shed.

There's an intimacy to this tour, and a warmth as Deborah offers
to reveal her personal favorite specimen in the museum: a rare, red
butterfly. G. W. Hinckley, it turns out, had a fondness for organs—
not the anatomical sort but the instrumental type, so that, amid
the taxidermy, the halls fill with sudden mirth: demonstrating the
workings of a reed organ, a volunteer pumps out some bars from
the wedding march while Deborah asks the group to picture, how,
when darkness fell, the music would be lit with its own candelabra
set upon a stand made just for that purpose that swings out from the
instrument on an accordion-like spring.

Deborah's voice is a trail in many senses of the word now, what
with the soft wheezing of the reed organ's wedding march, and I let
its lingering notes carry me even as they fade the closer I come to a
recrossing of the threshold of Hubbard's mood rooms. Though I'm
notebook-clad and penned, I'm without words on this visit: Caren's
absence has left a hole I do not seek to fill; if the dioramas envelope me, they resist articulation, so I give myself over to silence. For
this moment, I don't want to find words for them. Had I ever really
wanted to describe them? Not really. I only ever wanted to excurse
toward them.

Today they are accompanied by a playful, earnest whispering, of
minds applied expectantly to a problem of an afternoon. Two members of the group struggle in dim light to decipher mounds of piles
of handwritten letters that Deborah has brought out for them to
consider from the archives. It's a mammoth task that could benefit
from the work of volunteers even if the only thing they did was to
determine the gist of some of the correspondence between one Walter Smith and G. W. Hinckley. "Walter wrote every single day," Deborah's voice suddenly becomes apparent at a volunteer's shoulder,
"had he lived in our day, he'd no doubt would be a blogger." Then she
shoots across to me, "Do you think there's a mood to indecipherable
handwriting?" The bird room is as good as any for reading letters, I
suppose, and especially indecipherable ones.

"Mom!" a boy exclaims, but there's no proper punctuation for
the tone of his beckoning, spiked and rising like a question, immedi-

ate as a command. Always in these settings voices of adults comingle with their children's. "Mom! Oh my god! I found a loon!" Has the mother heard him? She seems preoccupied with decorum, with behavior, with disturbance, with keeping the child inside the reign of a familial hold: "Stop trying to look, and just stand there!" she says. It's a sort of richly inflected damper of a requirement no fiction writer could make up; it's a statement as real as the world can be. As a proverb, it might hold some wisdom; it could be a koan worth contemplating, and hadn't I already proposed that in order to see, we must be made to stop looking? As a dictum, it's just the opposite, sure to sound the knell for curiosity. "The peacock died," the other, younger, son observes, and "Hold my hand. I'm scared!" I lean to hear something the older brother wants to tell me about what he sees, but the mother yanks him back, "I would like you to stop talking to people and let them enjoy the museum!"

"Oh, really, it's no problem, he's not bothering me at all," I start to say, but she has already moved the children out of a Hubbardian zone of what they might think into the realm of what they should think. "Send it to Daddy and make him smile at work," she instructs the boys about something beyond my view. "Don't you want to make him smile every day?" I hear her ask.

If children could sound, what would they sound in rooms as these? "Hold my hand! I'm afraid!" I might have told myself on a visit to New York's American Museum of Natural History, my head no longer the house that holds the mind but a crash helmet battered by the noise of all the city's school children released at once as if from the zoo that more regularly contains them. There was no place in the museum to escape the grinding cacophony, and I found myself literally trapped in the hallway of a balcony in the Akeley Hall of African Mammals by an oncoming stampede of screaming mimis. Having run for my life, I didn't try to look but just stood before the famed gorilla diorama where I was treated to a dose of adult snideness: "Huh! I ain't related to no chimp!" one slumping man of a group of three asserts as he and his cohorts briefly regard the display before moving on: "Heh! Dat one looks like he's eating marijuana," a

woman says, before the original man picks up the thread of his earlier argument: "Uh! It ain't no chimp it's a chump. I feel no fucking
kinship with these goons."

All natural history museums are, of course, bedeviled by affilia-
tions and distinctions: I am not that animal. I am not that man. The
glass dividing us tells me that, but every time I've been to the L. C.
Bates Museum, I overhear a visitor who is literally in search of a trace
of their relatives. Standing in the presence of grandfather Hemingway's marlin is one thing; hoping to find mention of your grandfather, who, as an orphan, attended the school is another. Or is it? On
my follow-up visit, a young couple sits for hours in a set of wooden
folding chairs combing through Hinckley's *Good Will Record*, hoping
to discover the trace of their ancestor's existence there. In another
room, piled high with scatter and old tomes, I perform a species of
accidental research when I happen upon a mention in the *Record* by
the same Walter Smith whose letters the volunteers were attempting
to decipher. "I think I found your guy!" I yell into the Hubbard bird
room, excited to share the bit of novel information that's emerged. It
turns out that Smith had built a cabin near to the campus that he had
insulated with seaweed, thus prompting Hinckley to write an article
about what a brilliantly efficient weatherizing material seaweed is
and wondering why it hasn't entered into common use.

"What's worth remembering and who will archive it" weren't
questions the makeshift community of which I'd suddenly become
a part was consciously asking. The museum itself, its swale of voices
overheard, its quiet clamor, its incitement of interest, its unencum-
bered pull drew us into a zone that solitary researchers recognize as
the mood of the archive. Yes, it was the mood of the archive, and a
desire for relatedness that led me to begin to understand the L. C.
Bates Museum as a place for people in search of a different kind of
relative than the securely given familial one. Hubbard's mood rooms
and the museum overall house something that we can't find anywhere else and that we, in any case, need.

"The air is purer here," Deborah Staber says with the tacit assur-
ance of the obvious as if she were telling me to forget everything I

thought I knew and allow for the fact of the matter: cherry pie is red. As if to say, "breathe in"; it's what called Hinckley to this particular locale and its tract of land, she explains on a walk to her car for a tour of highpoints of the campus: "The air is purer here."

In his late-breaking autobiography, Hinckley recounts a defining anecdote for a personality and a lifetime. He tells of how one of his bachelor uncles, a prominent tailor in the town, who was also a follower of phrenology—the then-fashionable reading of bumps on the head to interpret character—suggested to Hinckley's parents that they solve the question of whether the boy should remain on the farm or pursue an education via consultation with the leading reader of cranial rifts and valleys, Professor O. S. Fowler, who just happened to be receiving customers in New Haven, which was just sixteen miles away by buggy and horse. Hinckley's family agrees to a consultation with Fowler for the cheapest option he has to offer: three dollars for a verbal report on their boy only, with permission to take notes during the session. Hinckley recalls Fowler's reigning pronouncement as though he had been that day in receipt of a tattoo, a laying on of hands that lay bare the mildly humorous truth of his being: "Young man, you're green," Fowler said. "You are very green. You are a regular Rhode Island Greening," the Rhode Island Greening being an apple that was popular in Southern New England at the time. "You won't be ripe 'til forty," he added, "you will improve 'til sixty."

Hinckley, a designated late bloomer, was the perfect figure for creating bridges with Hubbard on which adults and children could together pause rather than too quickly cross. Their museum is for people eager to suspend the tenets of developmental time and resist the conditions that hem the already "grown up." Still to this day, if the mood rooms suspend time, the museum itself seems suspended in time, as one recent newsletter attests. This month, a local student at the newly conceived alternative high school that now resides there is "working to help repair and stain the picnic table behind the museum. Come enjoy a summer picnic and a walk on the trails," the newsletter suggests. "It is field trip season!" the newsletter exclaims.

"The museum has been filled with happy voices of children enjoying exhibits. They have been delightedly catching frogs in the pond, walking the forest trails with our staff and making art inspired by nature." Local specimens continue to arrive at the museum in the form of donations just as they did in Hinckley and Hubbard's day, and this month "a mastodon tooth fossil that was dragged up off the coast by a scallop fisherman was donated to the museum by his family."

The motto of the museum is Inspiring Wonder, and this week both adults and children are invited to "explore the delightful small world of magical Winged Fairies and make a miniature fairy home, necklace or jar." "Come make a fairy jar, walk along a fairy trail, and see an exhibit of captured fairies or make a fairy house out of items found on the trail," the invitation reads, and then relays a slender thread of child/adult categories for each activity: "At 1 pm we are making stepping stones for fairies (all ages); 2 pm a fairy door (all ages); and at 3 pm making your own captured Fairies jar (adults and older children)."

My attempts to photograph Hubbard's mood rooms on my return visit to the L. C. Bates Museum are a disaster. Nothing in their purview is what one could call "clear," and the suite of snapshots is rife with bungling. How did a shot of nothing but my misshapen feet in out-of-date sandals come to be? Or a close-up of my own thumb and forefinger that looks like two globs on either side of an ass crack? What had I been trying to picture when the partially unzipped zipper of my pants interrupted the view? All of the photos seem smeared with the peanut butter of the elementary school–style sandwich I brought for lunch; most have the effect of what were called in the olden days "double exposures."

If I were to create my own invitation for the L. C. Bates newsletter, it would read something like this:

Leave your cameras and your words behind today as we, together, attempt to create our own moods out of a species of 'daub.' Keep an eye out for widow's peaks and beaks; for placidity, pose, and poise; furred faces and a blanketing of snow; streaks of shadow

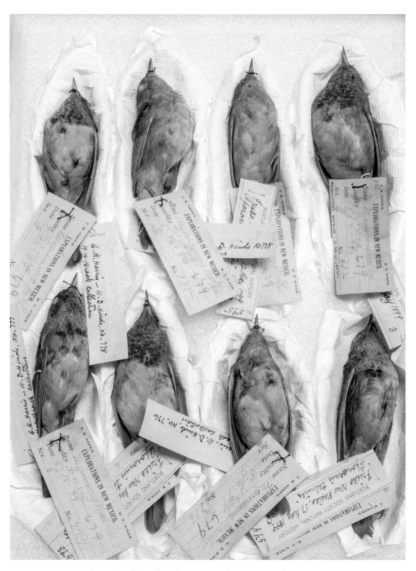

(1. *the effect of a reliquary made especially for saints*)

(2. *not to avert decay but to stylize it*)

(3. *some mood rooms recede while others compete for full occupancy*)

(4. *we are creatures of temperament, temperature, and tempos*)

(5. *taxidermists come in many stripes*)

(6. *to each bird, its leaf or tree; to each painter, his brushstroke*)

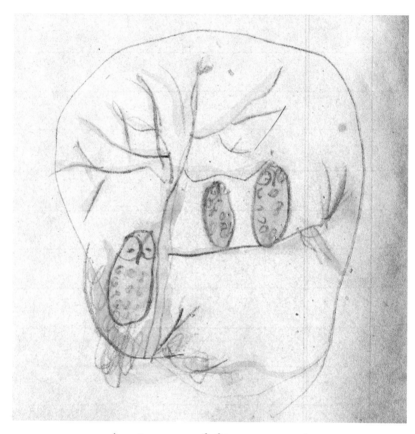

(7. *a transparency of a future present tense*)

(8. *the sound of a furred fluttering inside the walls*)

and wood-stained frames. As you prepare to enter the rooms of our habitat dioramas, especially the one in which the birds reside, see if you can catch the hint of something more than the bird before you: remember that there are always other birds residing in a separate sphere. Look out for the spirit of a "forgotten impressionist painter," who liked it that way, so don't try to remember him too exactly. Imagine the sets he painted for plays and pageants, in which case, in place of the birds you see here, stilled, or moving, picture a boy, orating, singing, bedecked and passing by. Allow yourself the luxury of ghostedness and shifting shapes: When, to your mind, do these mood creations melt a frozen form, or still a mobile one? Come meet the mood that's never left this place: it cups its hands to make binoculars; it bluffs its ears to draw nigh the whining of the wind inside a tree; it paints a door opening into a deeper space.

picture books

Every subject awaiting our inquiry must have an unhatched mood built inside it, hopeful for the entry of another consciousness into its precincts so that it can be birthed. I'm at a juncture in my mood trek where the more familiar I am with the territory—for a long linger of a time, the mood room menageries of Florence Thomas and Charles Hubbard—the stranger those rooms become. Sidling up to worlds made by people who are no more, discovering a trace of an imagination lost in space, is like finding oneself on a ship that passes another ship in the vast night of Being. Of course there is fog and the faint sound of a horn tuned to middle C when the comingling of one's consciousness with that of some fine stranger no longer living but alive creates a cojoined mood.

Let us enter a very particular type of room—a room lined with books and low tables—a local library's children's reading room as urspace of our own ontology of mood. It should come as no surprise, the further we fall down a Florence Thomas/Charles Hubbard rabbit hole, to find their own work as signal and flag, gracing the walls of just such rooms of absentmindedly kept-on winter coats and brightly patterned picture books. Charles Hubbard's oblong mural *Imagination* still hangs inside the children's reading room of East Haven, Connecticut's Hagaman Memorial Library for which it was commissioned, while a finely chiseled plaster bas-relief of Dodgson's Alice and her charges in Wonderland by Florence Thomas tints the surround of the Beverly Cleary Children's Library in her native Portland, Oregon.

Hubbard's painting, *Imagination* (fig. 27), has a companion piece called *Research* (fig. 28) that hangs in the original adult reading room, but, as ever, it's the suspension bridge between the two he aims for us to cross, explaining how he conceived the pair: "There are books which are the result of research and books which are the product of imagination, and books there are which result from the blending of these two." I like to think the emphatic diction of his *"books there are"* is his way of privileging the type that blends the two. We're in a library, but his figure for *Research*—the apparently adult theme—is a person seated at a desk out of doors, the place where we expect to find children at play, whereas *Imagination* features children so deeply seated inside the pages of a storybook, they seem themselves to be its illustrations. The man of *Research* looks up and out, but *Imagination*'s boy looks down into a book whose figures play inside a vast interior with room enough for sorcerers and saints—in this case, Circe and Saint Nicholas. The boy is studying and figuring; the man is gazing. Which is research, then? And which imagination?

Mightn't our earliest scenes of reading have been the place where we did our most fervent *studying*, our figuring and puzzling? We craned to know then, we couldn't yet discern, we were busy with *research* but attached by a finger, a yawn, a lock of hair to the mood of a picture, the incomprehension of a letter inside a word. As adults who claim to research, we're really staring into space, in search of the source of our childhood imaginings—that earliest study, both verb and noun. The mood of our mature searches springs from a picture—to each his accidental own—that once held us in a children's book, inside a room inside a library possibly tinged with illustrated walls.

~~~~~~~~~~

If Florence Thomas's Alice was on the wall, I'm not sure what would have emerged for me (fig. 29). How it would moor the mood of earliest reading. As an adult, I identify with the sign of Thomas's hand, the care proposed by her having sculpted in a mane for the rocking

(27. *room enough for sorcerers and saints*)

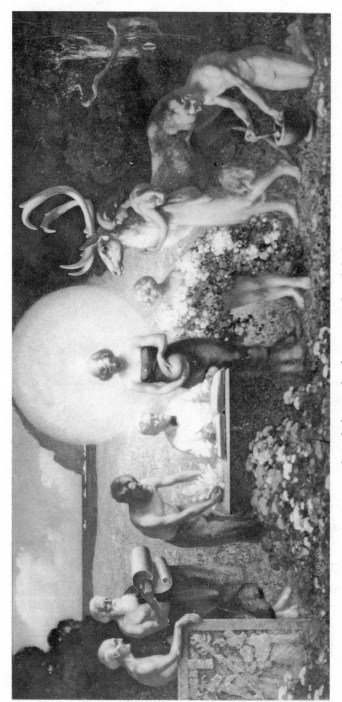

(28. *the hazy realm of sensuous nonknowledge*)

(29. *in a room lined with books and low tables*)

horse, of her arraying the figures in Wonderland against foliage and forest. I mistake her medium for Sculpey—her fingers are so live here—and wonder whose face she used to fashion Alice's from: if not for the flowing tresses, it could double as the face of a boy. I appreciate Thomas's eye for design: the way the triangles carved inside of Alice's apron quote the shapes that line the arch that holds the scene together. I feel inexplicably moved by the face encased in a flower because, in lieu of an easier and insipid smiley face, she has lent it an expression that makes it seem "human" (fig. 30).

If, when a child, I had been greeted by Florence's forms, I think I would have confused the pillows of plaster with cookie dough and have wanted to eat them. I'd have been rapt by a particular detail, to which, in each visit to the reading room I'd return, wondering what Alice might be hiding in her hands. Other details might have frightened me: What was the mood of Alice's face? Was it sullen? Was it stern? Why were her eyes like cavities? The girl commands this space—the Rabbit and the King of Hearts do her bidding by her feet—but, no: the space is really lorded over by a floating cat's head with a bedeviled, broad-faced grin.

That a Hubbard-Thomas path would eventually converge on the walls of a library's reading room for children made all the sense in the world to me, one east, one west, on our nation's compass. Less clear to me was a midlife wending that had me arriving on the doorstep of a set of books written for children. It was as though one day I awoke with an appetite for "a good children's book," in the way one feels a need for a good mystery, a good meal, a good night's sleep, or a good spanking. Forget Nietzsche and Proust and Thomas Burton as the best most obvious tutors on a syllabus of mood. The answer, my friend, is blowin' in the wind of the picture books of one Margaret Wise Brown.

~~~~~~~~

Unlike most people, with or without children of their own, I had not come to Margaret Wise Brown by the popular-to-the-point-of-ubiquitous *Goodnight Moon*. Even if I had never read *Goodnight Moon* (but hasn't everyone read *Goodnight Moon*?), it was impossible not to encounter it in every children's bookstore and airport across the land where it not only existed in its original form but in endless imitation from *Goodnight* ——— (fill in the blank of the city or state

(30. *how it would moor the mood of earliest reading*)

you are in) to *Good Night Farm, Zoo, North Pole, Ocean, World, Galaxy*, and *Baby Jesus*, to say nothing of parodic spin-offs like *Goodnight Goon, Goodnight iPad*, and *Goodnight Husband, Goodnight Wife*.

But *Goodnight Moon* wouldn't prove to be the mood lure that would have me wanting to enter more deeply into the worlds created by Margaret Wise Brown. I found Brown lurking with a start in the pages of a nine hundred–page book on noise by cultural historian Hillel Schwartz. I had invited the scholar and translator to the campus where I teach to discuss his book in a class in "literary acoustics"—a course in which we sought to transform words on pages and so much of what we took about our lives to be self-evident by turning our attention to sound, voice, noise, and silence. Schwartz introduced the students and me to one book in a series of books devoted to listening by Brown, the paradoxically titled *Quiet Noisy Book*.

It was interesting at the outset to think about the vast mood influence the magic of one of Brown's books had cast into the nighttime wells of millions of children over a period of several decades and still to this day. Then, to pause to consider how little any reader, be they parent or child, knew about the particular geometry of her life, to say nothing of the scores of books she wrote that haven't yet enjoyed the same ascendency as *Goodnight Moon* including her *Noisy* book series, or those she wrote under a handful of pseudonyms. Could it matter to our experience of the book to know that Brown didn't live to see *Goodnight Moon* thrive, that she died young, at forty-two in 1952, exiting life with the kind of boisterous exuberance she was known for: cause of death was a cancan-type kick of her leg into the air following a minor surgery. She died instantly of an embolism. In an equally strange twist of fate, in her will, Brown had named the child of a friend the right to all monies earned by her books should he survive her, but the boy, who never completed high school and who gained a reputation for destroying public property and beating people up, grew up to squander the millions.

At the center of the famous tale was a bunny, and Brown had grown up with bunnies and other animals. As an adult, she once took a bunny on a train with her en route to a date; but, could a reader imagine it?—as a child, she had also skinned a bunny. The houses she lived in were like hutches—in the middle of Manhattan, a cottage that she called Cobble Court; on a ruggedly beautiful island in Maine named Vinalhaven, a former quarryman's house she dubbed "The Only House." Brown's biographer, Leonard Marcus, describes the house as an uncomfortable extravagance: it wasn't wired for telephone service or electricity and lacked indoor plumbing, but Brown made sure the larder was stocked with "champagne, fresh cream, imported cheeses and other such necessities." A well served as a fridge, as did the cool, running waters of a nearby brook, from which she might surprise a visitor with a bottle of wine. Rainwater was collected for bathing, and three outhouses were positioned "for the sake of the view": "a mirror had been nailed to one of the apple trees in the yard, and a pitcher and basin were left out on a battered Victorian washstand. First-time visitors would be surprised on opening the drawers to find the freshly laid supply of scented soaps and toiletries of Margaret's 'Boudoir.'" One part of the house brought the outside in through a collection of small mirrors, "each differently framed and made flush with its neighbors." The mirrors were hung across from a door that opened onto a "sheer fifty foot drop," and positioned in such a way each to reflect a "different image of the sea" in succession. People finding themselves in the mirrors, would glimpse themselves as "plural."

I don't recall which detail from Hillel Schwartz's work on Brown and noise made me ask him about her romantic life. I only remember a feeling of a life lived differently and athwart, and a desire to draw closer to her, whatever the answer might be. I wasn't looking to find myself in her, but learning that she'd spent most of her life in a relationship with a woman poet who called herself Michael Strange and who had been the ex-wife of John Barrymore filled me with a perverse delight. Picture this: when Michael was in the mood, she called Margaret "Bun"; when Margaret was in the mood, she called Michael "Rabbit" (those really were their pet names for one another). I loved how the idea of a "childless" lesbian having devised one of the most classic books for children gave the lie to maternity as the only or most natural route to knowing how to be with children. I wondered if hospitals that sent copies of the book home with newborns would discontinue the practice if they knew the book's author had been queer. Would all those folks who lulled their children to sleep with *Goodnight Moon* rest easy if they knew the little prayer was birthed by a lesbian consciousness? Might they worry that *Goodnight Moon* could unconsciously shape a mood realm deep inside the child that might make him grow up gay?

Come to think of it, all of the mood-room makers to have drawn my interest are by coincidence queer insofar as all three—Florence Thomas, Charles Hubbard, Margaret Wise Brown—lived out of tune with the hum of a neatly domesticating hetero norm. Florence Thomas, I recently learned, and as I'd imagined, enjoyed her days in retirement alone with her cat, her garden, and a basement studio where she carved abstract art out of wood. "'I'm a cat person,' Florence Thomas told a local reporter in 1971, 'So I am really fondest of 'Puss and Boots.'" "You have to know how to be a scavenger," she said, referring to the suite of materials used in her scenes. "It's much easier now since we have plastic flowers and leaves. I used to bring live things from the garden." Though retirement would give her time "to renew old friendships"—she was planning a major trip to Australia and New Zealand with one friend—she explained she wanted to work on her garden first. This wasn't a garden-variety garden providing flowers for the table or tomatoes for a stew. Spanning one half-acre, it was a form of applied dedication by a specialist-eccentric: Thomas had cultivated one hundred varieties of rhododendrons. She continued to make 3-D constructions, but now by way of what was known as the *Personal* View-Master camera: a device that people could use to make View-Master reels with images they themselves took, based, that is to say, on their own lives.

Hubbard wasn't exactly a solitary soul, but he lived at an apparent distance from his wife, who has been described as emotionally unwell, devoting himself to his students, his art, and his summer travel in turn.

Brown had experienced attachments to other women and a near-marriage to a man before she met Michael Strange. Unfortunately, her relationship with Strange wasn't, in the long term, a happy one. Strange comes off as abusive in the end, with Brown the ever-pining and unfulfilled lover. One of numerous painfully poignant junctures in Leonard Marcus's account of their relationship has Brown hiring a lobsterman to build a special house just for Michael Strange near to Brown's "Only House" on Vinalhaven. Since Strange rarely joined her there, perhaps Brown thought designing a "Picture Window" using an ornate picture frame she'd found in a Rockland antique shop trained on a spruce forest would do the trick. But she was wrong. Nevertheless, an unconventionally attuned Brown was involved in experiments in living and in writing all her short life—having been the literal voice to prompt Gertrude Stein to compose a children's book, with the result being Stein's 1939 *The World Is Round* illustrated by the same artist who pictured *Goodnight Moon*, Clement Hurd. Brown's *Noisy* book series, which came into print around the same time, grew directly out of her work with Lucy Sprague Mitchell, founder of the Bank Street School whose experiments in early childhood education in the early twentieth century included encouraging children to be noisy rather than disciplining them into silence. Many of Brown's books in progress were tried out, so to speak, on children at the school.

The *Noisy* books, Brown explained, "'came right from the children themselves—from listening to them, watching them, and letting them into the story when they are much too young to sit without a word for the length of time it takes to read a full-length story to them. I mean three- and two-year olds." At the center of the series is a little black dog named Muffin who stands in place of a child—one imagines the child-reader identifying with him—but who, on account of his being a dog, has a keener sense of hearing than a child. Or does he? The books are meant to meet children in that time before they've come to use language fully to mediate desire and their world, that period when looking hasn't yet replaced listening as the form of reading, where letters aren't yet yoked to sounds but float, one form among others on a page splashed with Leonard Weis-

gard's bright colors and Brown's colocation not just of words bound
to meanings but of phonemes' clattering in imitation of sound.

If Gertrude Stein famously discovered the difference between
sentences and paragraphs by listening to the rhythm with which
her dog lapped the water in his bowl, Margaret Wise Brown understood the equation in reverse when she included dogs among
her listener-readers. A dog can train a writer to make prose; and, a
writer can write a book that a dog can understand. "Dogs," Brown
wrote, "will also be interested in 'the Noisy Books' the first time they
hear them read with any convincing suddenness and variety of whistles, squeaks, hisses, thuds, and sudden silences following an unexpected BANG."

Here I am, a dogless adult, tethered to and transfixed by Brown's
protagonist dog. The play-space she creates for him and invites us
to enter in her books isn't exactly a-brim with frolic and jaunt, a
run in the park, a pat or a pant, the tossing of a ball before being
tucked in with a biscuit and warm milk for the night. These aren't
easy or straightforwardly narrated books. They are tantalizingly difficult; most have the riddle of sound at their center; all are marked
by absence, for isn't it the nature of sound to emanate from the object world but remain detached from it; to seem to reside within
an object but not be traceable to it; to exist as a solid experience of
sense but to remain ungraspable, sometimes unlocatable, at heart,
ephemeral?

Meaningful and meaningless sounds were the bases of our
moods—possibly the earliest molds for the form a mood could
take—and sound is mood's analog.

~~~~~~~~~~

In the *Quiet Noisy Book*, Muffin is awakened by a "very quiet noise."
"What could it be?" the narrator asks, then provides a set of answers
in the form of questions whose examples are beyond audition, or
imaginary: "Was it butter melting?" "Was it an elephant tip-toeing
down a stairs?" "Was it a skyscraper scraping the sky?" Teasing the

limit of our senses to meet the reach of our imagination, gathering into a book-as-net all that coexists simultaneously with us and beyond us, offering a promise of presence—the blank filled in—if only we could be willing to listen hard enough, the book answers each possibility with the word "no," without at the same time canceling its belief in the possibility that what Muffin *could* be hearing was "a fish breathing," say.

What do we wake to? What is the sound that's waking Muffin up? It's something very quiet, but there. It's "quiet as a chair. Quiet as air." It's something Muffin knows, but do you know what it is, dear reader? By now you could be terrified or mesmerized or both, when, splash, it is the sun coming up. The sound of the sun coming up is accompanied by conventional harbingers of joy—roosters crowing and such—but Brown also leaves the child—and the adult reader— with puffs of cloud heavy with mystery, my favorites being, it was the sound "of a balloon about to pop," and "it was a man about to think."

Here Brown had gifted me the finest definition of a mood, and one defying explanation. Mood: *it was the sound of a person about to think*. What qualities inhere in the sound of a person *about to think*? Does thought really admit of pause? What form of silence have we here, for doesn't the "about to," though yoked to sound, imply an absence of sound? A soundless sound?

Philosophers of silence spoke of silence as its own active presence, not merely a negation of sound. Of brusque or smooth silences, of silences yoked to utterance, and requisite to language's meaning-making capacity, and silence as its own live phenomenon, with its own temporality, independent of sound. Of silence as predictive of the type of utterance to follow, of rhythmic silence, silence as the place of anticipatory alertness, and silence that serves as an end point or terminus, a concluding silence. Silence that casts a spell, and silence that lifts a spell. Certainly, there is no such thing as silence pure and simple, and if any of us were asked to generate types of silence, we could come up with enough examples in an instant to fill a page—just now, radio silence; the silence enjoyed between intimates; awkward silence; and how about the silence that pervades the experience of reading? Does that bear any resemblance to the sound of a person about to think?

Margaret Wise Brown is the person whose picture books ask these questions, not I, and when she translates the "new day" in *The Quiet Noisy Book* into the dawn of life's unanswerables, she opens her reader to possibilities of heaviness or light, of greeting the unknown with a mood of curiosity or hiding under the covers depressed by the prospect of pondering.

If we equate the sound of a person about to think with a form of soundlessness, that's only because, as adult readers, we think of thought as word-bound, a kind of talking to oneself, and sound, therefore, as one sign in a system of language, like the space between words, say. But what if the sound of a person about to think were a tone or air, a vibration or an atmosphere that thought depended on, no matter the form thought took; what if it was understood to be a presence rather than an absence, albeit one we might not be able to translate into words?

What I notice about the books in Margaret Wise Brown's *Noisy* series is how, in each of them, she confides in absence, allows for it, enfolds a reader in it even: she makes absent presences a field of play. For, whether it is *The Indoor Noisy Book*, *The Noisy Book*, *The Winter Noisy Book*, *The Seashore Noisy Book*, *The Summer Noisy Book*, or *The Country Noisy Book*, to name a few, there is always something menacing, askew, uncanny, inexplicable, in excess, or ghosted to contend with or consort with. It's that same quality that provides a mood for or to childhood but that is forgotten or covered over in adulthood, in which case, adult moods are a form of the comforting blanket that one had lain beneath when such stories had been read to one; they are the cover for a more mystifying cloudscape forgotten but ever-hovering.

There is a kind of ultimate silence, of course, that, if we try to "think" it, fills the mind with existential dread—the blotting of consciousness, the death-in-store that would jettison us into, well, I can't think of a better figure for it than "outer space." Celestial music or the music of the spheres must just be something humans have devised to console themselves with when confronted with the inconceivable absoluteness of the silence that is death. Mood is that silence momentarily animated.

~~~~~~~~~

In several of the *Noisy* books, Muffin's eyesight or some other aspect of his sensorium is compromised so that he has to rely more fully on his capacity to listen. In *The Noisy Book*, he's gotten a cinder in his eye and is required to wear a blindfold. In *The Indoor Noisy* book, he has caught a cold and is forced to stay inside making him more acutely aware of the noises in the house. In *The Seashore Noisy Book*, he is disoriented by the fact of his hearing, for the first time, the sounds of the ocean. As the book nears its end, he hears a splashing sound, and, as always, we are invited along with Muffin to try to attach the sound to its source: "Was it the sun falling out of the sky?" "Was it a sea horse galloping?" Earlier, we're asked to listen for the light of moon and stars on water. The "answer" to the splashing is that Muffin has fallen out of the boat and it is his own flailing that he's caught in the sound of. He began by falling into a mood—the mood of the sea—and ends by being rescued from his nearly drowning there. This particular meta-ending is just one instance of the mystery of sound—and by affiliation, mood—being answered by a scenario of oddly proportioned dimensions in which nothing is resolved but where we are asked in effect to squeeze the large round ball attached to a scary clown's nose.

The end of *The Noisy Book* might be the most delightfully perverse in this respect. Here, blindfolded Muffin hears a squeaking whose source he can't identify. Neither a garbage can nor the house nor a mouse nor a policeman, the answer is "a BABY DOLL / And they gave the baby doll to Muffin for his very own." The problem, or pleasure, for a reader, though, is that the anthropomorphizing has Muffin playing with a facsimile of a human rather than with a doll in the form of a diminutive dog—a puppy doll, and that the illustration shows the doll to be nearly twice Muffin's size. He can barely hold the doll, but

(31. *suffused with a smell of sour milk and hay*)

he's propped himself, sans blindfold now, as if posing for a snapshot.
Muffin and the reader in their collaborative investigations of sound
and mood seem to have won the booby prize (fig. 31).

The Winter Noisy Book might broach the most uncanny territory
of all once we realize it's Brown's version of Muffin Had Two Daddies.

Winter has come; night is cold, and still, beyond the windows; stark, black branches rattle against the windowpane. Muffin hears a thud thud thud scrunch scrunch scrunch—what was that? Rather than provide her usual series of playful speculations, Brown works together with her illustrator—this time an abstract painter other than Weisgard, Charles G. Shaw—to produce a frighteningly stark page. The illustration shows an empty dog collar and leash, and a shade drawn against the darkness in a furniture-less room. One sentence is spelled across the white floor of the room. As answer to the question of the scrunching, it could describe the entry of a sci-fi menace: "The fathers were coming home" (fig. 32). Does the empty collar mean the dog has fled or that the dog has died? More benignly, could it mean the fathers are the ones who will take Muffin out for a walk? The ominous mood is dispelled—the blank of the page filled in—by the introduction of a perfectly normal-seeming domestic scene comprising two men. They produce sounds that Muffin listens to, like laughter and the dropping of a nickel, the turning on of a light, and the b-r-r-r-r of a shiver before they stretch their legs in front of the roaring fireplace where they sip cocktails and crunch on celery. Muffin is merrily spread on the rug before his bachelor fathers—one clad in neatly striped pants, the other, solid blue (fig. 33).

Leonard Weisgard, who had been a dancer and a Macy's window display designer before he turned to children's book illustration, mastered the magic of minimalism in the Noisy books. The palette is simple, relying sometimes on no more than four colors in the course of a book, including the noncolors, black and white, especially in combination with primaries. A page could turn entirely yellow, daylight reduced to a trapezoid of blue against red. He arrives at images—the shape of icicles, for example—by subtracting rather than filling in, or by cutting out and pasting in. If I find myself wanting to roam around inside an illustration, is that the same as reading?

The endpapers of The Indoor Noisy Book hold my attention the way the interior of a doll's house once had. For a two-page spread, he's cut one side of the walls out of Muffin's four-story house so we

The fathers were coming home.

(32. *it matters that I touch the page when turning it*)

(33. *these aren't easy or straightforwardly narrated books*)

can climb the small ladder in the basement through a hatch to arrive in the dining room whose table sits below a Victorian tub in
the bathroom on the floor above it, which sits beneath the most interesting room of all: the attic. There's a bed frame and an equally
empty, though ornate, picture frame. There's a dress frame with ex-
aggerated curves. A bandbox and a dome-lidded trunk. A taxidermied moose head. There's a piano.

Is the power of the drawing—see how it draws me even now, drawing my adult eye, compelling me to redraw it by way of words—fueled by nothing more than reminiscence—just one more instance of a version of 3-D paper cutout houses enjoyed in childhood? When a mood space is activated by a work like this, we are put in contact with something ever present but impalpable. It's never so simple a thing as a reminder of or a return to something from our past; it's more a *how* than a *what*.

There was a sentence like a conundrum that appeared in Hans Ulrich Gumbrecht's work on mood that I also wished to understand and most definitely felt I was experiencing. Describing what happens when literature creates a mood, Gumbrecht made a distinction thus: "For what affects us in the act of reading involves the present of the past in substance—not the sign of the past or its representation." I'm not compelled by Weisgard's illustrations because they bring my past back to me by pointing to it in replica but because something has been made possible in the readerly transaction of color, line, and consciousness that puts me literally in contact with a piece of a reconstituted atmosphere as if to say I once was in that dream but forgot I'd never left its precincts. Now I not only recognize its outline but also feel what it is to be buoyed inside its envelope. It's not a forgotten place but a place where I always am—part of my mood repertoire—but which I fail to see.

While Weisgard's elaborated interior might seem like just one more iteration of a repetition of examples—enter a return to an already-scripted View-Master—there were other aspects of his illustrations for Brown's books that turned a key to unlock a mood and that were much less legible. They were not at all "representational." These were absorptive planes (pure blots of color, where the feeling of the paper as recipient of a hue was as important as the fact of color itself), and stencils.

Numerous of the images in *The Indoor Noisy Book* are finely dotted or shaded as though they were applied through screens or with sponges onto the white page below rather than fully filled in, and bunches of tiny flecks of color mist around their edges like the scat-

ter left by a spray-painted stencil. Maybe a child is meant to read
such touches as a sign of its maker having fun: "These pictures are
ones you too could make," the speckled fringes say, "some afternoon
of arts and crafts."

For this adult reader, it matters that I touch the page when turning it, and when I do, I'm suffused with a smell of sour milk and
hay. White-colored goo had been applied to a piece of bright red particleboard, four by four inches square. Had we been given sprayers
or had the adults sprayed our hands? You pressed into the board,
painted the white goo on, and what remained as an absence but
also as an imprint was your child hand. Overly warm milk in a small
waxen carton probably was meted out to us after nap time in kindergarten like dribbles of white into the mouths of mewing kittens. But
the "stuff" we used to "make our hands" that day was milky, too, and
sour smelling. The hand left on a block of red detached from a body
was meant as a gift to our mothers, but it wasn't something I wanted
to give to mine. The hand lay in a drawer as a shadow self. It was the
hand from a body that could only be described as "bleating." If this
wasn't the stuff of an originary mood, I don't know what else could
be. And it was immanent in the quiet noisy books of Margaret Wise
Brown.

Stencils could be a form for mood itself: there was always something missing from them; a stencil was a means, and its end. Would
the stenciling in *The Indoor Noisy Book* create a mood no matter the
particular reader's relationship to stencils, then? Or in order for it
to do its work, need it be tied to a remembrance of stencils past? To
me, it seemed struck from the same granules as a mood that I harbored, one part quavering voice, one part shadow.

~~~~~~~~~~

Stencils can be found in caves dating to 10,000 B.C. And guess what
they depict? Hands. Entire walls with nothing but hands like paw
prints as records of human yearning. Early man, delighted with a
technique of missing and finding himself at the same time. Hands

as outlines feeling for something in the rock. Not pushing it upward like Sisyphus but transmitting or receiving warmth from it. Then there were those uses of hands that magically—with no more than a set of stick legs here, a wattle there—transformed the outline of palm and fingers on a page into a turkey. I wonder if we effected such magical transformations in the same week that we spray-painted our hands or in the week following. No doubt we were also instructed to give our turkey-hands to our mothers.

In even earlier childhood, when naming is the game but not yet reading, you might be placed before the sort of "puzzle" made of wooden figures that you are meant to fit into shapes cut out for them. The shapes—either abstract or figural—have knobs attached to them that make them seem like lids or doors. No doubt the point of such puzzles is to aid a child in the development of hand-eye coordination while at the same time teaching her words. Assuming that is a skill I've mastered, what kind of pleasure could such a puzzle hold out for me now?

I bought one such puzzle at a flea market because I had a hunch it belonged with the View-Masters and picture books in a cupboard marked "the mystery of mood." I also found it aesthetically pleasing: in its world, a duck could reside in the same row as a clock, a house was the size of a crow, and a cat that of a sailboat. Each thing was unstuck from any context and returned to its thingness: one purely red chair; a single boot without the need for a foot or mate. I felt it had something to teach me, but I also quite simply enjoyed it: placing the wooden tree snug in the slot suited only for it and it alone, doing the same with the dog, the hen, and the jug must work to quell the chaos in any adult life.

The principle behind the puzzle was as riddled with absent presences as any book by Margaret Wise Brown and just as rife with a mood of earliest reading. Fitting each figure into its slot, you earn the prize of getting something right, case closed. But you also cover something over; each time you remove a form, something is revealed: hiding inside each home for a form is its word. It's easy to get the words wrong—one reason no doubt that the manufacturer called the puzzle *Simplex,* brother of complex: instead of "boat," the slot yields "yacht." Where I recognize a creamer, it tells me it's a "jug." Instead of a crow, it offers the word "bird." An obvious Christmas tree is simply a tree. The figure meant for a slot whose inner wall reads "teddy" is missing from this puzzle. I think if I saw it, I'd call it a bear.

Learning to call things by their right names; imagining words as entities, or answers, that hide inside things; trading in the feel, and the shape, of the thing for its word so that you can eventually advance from puzzles to reading: our moods lurk in the interstices, for our mood is the story no one asked us to devise when presented with the shape of the thing. It's the missing link between the word and the thing, and it's also the cloak that shrouds the word in mystery. It's the shape left inside the toy box cut off from the hand that held it.

It was the sound of a balloon about to pop.

~~~~~~~~~

Let us close our eyes and feel our way. Let us feel our way back to the land not of nod but of mood. Let's get back to basics.

No re-search is ever finished—that's why we call it research—and especially not one as boundlessly cavernous as mood. Just when I think I've completed my mood room re-searches, I wend my way back to a beginning. Mood quests have no end, but you've got to stop somewhere. Why not close at the place where our subjects began? Not in order to unfold yet one more tent but to fold a tent up and appreciate its compactness. Following the composition of these pages, I felt compelled to make a trip to the place where life for Hubbard and Hinckley so to speak began, to Guilford, Connecticut.

The story of that trip doesn't mean to unfurl. It's not a spool unwinding, but a box filled with essences. Were I to place them on a shelf in a room inside a children's book, they'd consist of one basket; a wounded soldier; a village green; a historian's hand; a librarian's smile; a spire; an attic; and numerous pies.

Of course, we find a youthful Hubbard bounding there, youthful even in old age. There are letters from former students who loved him so, and articles. In one reminiscence, a former art student from the Commercial High School in New Haven remembers that "although he was 63 years old in 1938, this fine teacher was as young, or younger than any member of his class. He had that particular magic some grown-ups have that make young people want to be with him. Every class was a new adventure."

Every year, she says, their art club (they called themselves The Demented Art Fakers Association) took a bus to Guilford where they met Hubbard for a picnic. "From then on it was 'shanks mare' through caves, fields, pastures, over hills and dales of Guilford. We climbed stone fences, waded little brooks, crawled through old caves on our bellies and tried to keep pace with our mentor as he pointed out many beautiful sights. . . . 'See the way the light falls on the rocks over yonder. . . . Look at the beautiful pine tree, can you see how the tree's shadow falls on the ground?' He never stopped teaching us."

Pies go with the attic, and the basket was attached to a bicycle. The Guilford Town Green was a defining feature of the town that Hubbard had re-created by interviewing octogenarians about the way it looked in Guilford's early days. Fronting one edge of the green: a classic white New England house of worship where Hinckley and Hubbard met. Along another side of the green, the house that bore the attic where Hubbard made his art and where he also invited his students, following their lessons, to pie parties. The basket also went along with a box because when I arrived in Guilford on a chilly November morning, I was greeted by members of the Keeping Society in a place called the Medad Stone Tavern where a cousin of Hubbard had one time lived and that featured a room devoted to Hubbard. One guide, a former principal of the high school who was dedicated to preserving Hubbard's story, met me with a box. Another guide kindly noted that age and weak knees made sitting preferable to standing, and she led us down a set of stairs to a large, round, accommodating table.

The pace of sifting through a memento box is its own mood of unrushed temporality, an adagio peculiar to each box's owner, each pause, a fore- or after-silence that punctuates each perusal of a thing therein, unscored. Sitting with someone who lets you sift with him is like having a date with a form of another person's silent reading. The things in the box are piled like layers of unbaked pie; the things slowly rise as if in baking it. The pace bears no resemblance to the huff and fury we've all witnessed in fathers who failed to read the proviso on a toy's box, "assembly required," though it does bear a relation to a tinkering with toys—more like the pace of an adult hobby, that moment circa 1971 when working-class fathers, aspiring to "dens," set up shop in basements where they took to making model boats in bottles, the key feature of which was the patience required to hoist the miniature rigging and sail.

A POW had once enjoyed the freedom of encampment. In a black-and-white photo that emerges from the box, he's roughing it, shorted and shirtless before a well-pitched tent, he's Hubbard's young companion of a scout in Maine, Dick Chapman, and the model for more than one illustration. When he returns from imprisonment in the Pacific during the Second World War, he's skin and bones, a mere eighty-six pounds, and he's never afterward the same. His mind is damaged.

The mother of my guide had owned a grocery store from which she sent him on a special errand circa 1952. Word was that a nephew of Charles Hubbard was clearing out Hubbard's attic studio, Hobgoblin Hall, and selling the leftover paintings that remained there. The mother took all the money from the grocery store till—about a hundred dollars—and sent the twelve-year-old boy on an errand to buy however many paintings he could secure. The store's bicycle had a large basket for stowing groceries, but on this day he rode through the streets balancing, there, the four paintings he had purchased from a very tall man who answered the door at Hubbard's house, *Logs on the Kennebec* and the *Great Basin of Mt. Katahdin* among them.

My sojourn inside the box makes me late for my appointment at the local library's Edith B. Nettleton Historical Room. In the library that has me returning to the green, boxes are replaced with a

closet packed with Charles Hubbard ephemera; a set of map drawers with same that we bend down and into; and glassed-in shelves
holding books that require the use of a step stool to reach. The librarian smiles when she opens the closet door—not a broad, beam-
ing, sentimental smile but a subtle, knowing smile. She knows I'll
be delighted and she appreciates the spirit of the difficulty of the
challenge of finding a *way* to search out what's important amid layers of different types of things. The room I'm reading in is named for
Edith B. Nettleton, who only just died in January at the age of 105.
She was still volunteering at the library one day a week in her advanced old age. To work in this library is in a sense to meet her, but
it's strange to know that my visit here has missed her only by months.
The walls of the anteroom are lined with fifteen of Hubbard's oil
paintings—cloud colors and blue receding mists predominate—
and when the librarian swings open the closet door, I experience the
rush of a memory of a different order: the day on which my brother
combined a promise and a threat, who, in giving me knowledge,
subtracted wonder without canceling question, when he opened the
door of the closet in my parents' bedroom to give the lie to Santa
Claus and show me where my mother hid the gifts. Amid the clothes
were toys and treats stowed in boxes piled high, visible now, but out
of reach.

What's important to one person can never be the same as what's
important to someone else, just as an object as expedient as a nail
clipper takes on an aura of secret interest to a reader in a future age. A
leather-bound six- by eight-inch book residing in one of the drawers
is just one of many things; it doesn't enjoy a special home or file, nor
should it, but I read it as the Thing from which all the other things
spring. The main thing worth keeping? It's a discarded ledger book
dating to 1832, some pages of which are filled with a meticulous register of sums and names, while the bulk of the blank pages double,
according to the book's inscription in a child's hand, as "Charlie
Dannie Hubbard's" "Drawing Book." Hubbard was born in 1876, so
this ledger book must have been a familial leftover, something from

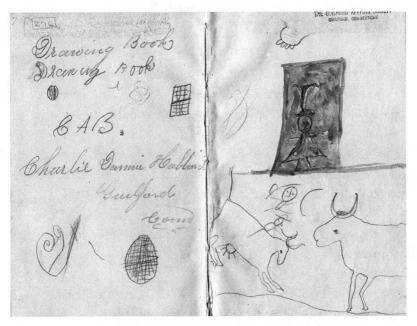

(34. *the page is a porous rather than a delineated plane*)

a familial archive that its owners found, if not worth keeping, at least available for repurposing.

The drawings in Charlie's diary-primer all rely on a combination of pencil and watercolor. There are traces of lessons and traces of play. If you tilt a half moon on its side, you can make an animal's horns, someone might have shown him. Here's an entirely blue woman in an entirely green doorway; the suddenness of an udder; a steer tilting downward to buck (fig. 34). Draw an egg, and now a spiral, because the world is made of ovals and curves as are the letters that form the words that spell your name. At what point did we forget this?—that letters are something we learn to draw, and pictures are something we learn to read.

Spokes, hoops, and seesaws. A bicycle whose seat rises high above the ground. The key that turns a kerosene lamp. A sill of potted plants. A green-handled brown-fabric umbrella. Numerous birds. Sometimes images exist unto themselves; other times, they are part

of a narrative. A figure with an overly large head and spiral eyes
could be a self-portrait cartoon. A man whose hat has popped off
and whose fists are raised in the air suggest a narrative. We can hear
someone offstage saying, "Put up your dukes!" Occasionally, words
in the form of sentences but without a sense of the spaces he will
one day learn to place between the words appear alongside an image as an early attempt at illustrated story. One tells of a cat trying
to protect her kittens; another is a dialogue in which a boy tries to
determine the capabilities of a blind companion. The cribbed words
hang inside lines like notes to a staff (figs. 35 and 36):

can you run yes I can run but I but I cannot play you may play
you may ride a

horse but I cannot run ride but yes play horse but I cannot see a
cat or ride a horse you cannot see?

Consider all that is allowed to share a page inside a child's eye.
When an adult juxtaposes like with unlike, we call it surrealism. But
for the child, the page is a porous rather than a delineated plane.
The book separates itself into pages, but it may as well be one endless edge of a boundlessly vast sheet. The page is a mood space upon
which a multitude of forms gathers and converges. On one ostensible
"page," Charlie draws a very large fish; an animal playing a trumpet;
a bush; and a house that is smaller than the fish but whose windows
are too many to count. If there comes a day when you cannot stop
counting windows, we consider you mad. For a child, primal pleasure is based in repetition, and making things that test and surpass
one's ability to count is tantalizing. A man wearing a derby appears
atop the house's chimney, but it's not clear if he is meant to emerge
from the house or if he exists apart from it. The figures are unrelated
to one another in their separate parts of the page at the same time
that Charlie has linked them by painting them all orange (fig. 37).
    Though, at the end of a day's work in the Historic Reading Room,
I find nothing that relates directly to Hubbard's creation of the

(35. *I only remember a feeling of a life lived differently*)

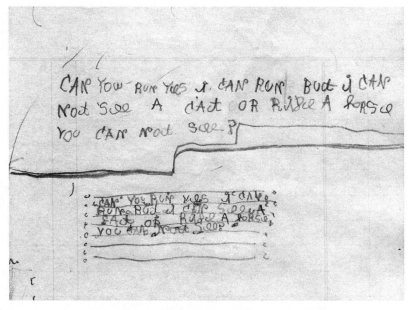

(36. *like notes to a staff*)

habitat dioramas in the L. C. Bates Museum, his childhood drawings read like blueprint prototypes for his future work, traces of the
absentminded wile that stoked the future preoccupation and vision.
Jokingly, I can decide that one childhood sketch of a bird appearing
to land atop a museum is prescient (fig. 38). The relationship between three owls in a tree of childhood imagining and the later work
of positioning full-scale owls onto branches into a scene that's been
researched could be as accidental as a mood (plates 7 and 8), but the
thing about his childhood sketchbooks—and later, his day or commonplace books—is that they are also records of the delivery of the
self to a regime of practice, training, and preparation for that which
one cannot know but to which one arrives ready with a skill.

    What do a teacup and handbell, a vase of flowers and a fork and
knife, a treble clef and a skeleton key have in common (fig. 39)? Differently colored and hovering, with no lines to connect their dots,
suspended between representational and non-, they create a mood,
of *what*, I cannot say. They are evidence of a child mind working on

(37. *consider all that is allowed to share a page*)

(38. *a desire to draw*)

that puzzle again of fitting objects to their slots; of matching each thing up with its proper place or space in the world; grouping each thing with the thing it is like. But the way they hang in their own unlined but idiosyncratically ordered space, nodding and tilting and curving all the while "arranged" in yellow, red, and green creates a kind of score, as if in music, and a sort of personal cryptogram: a mood.

Both more and less than a system of signs, the images don't directly translate; they *are*. Are they merely traces of practice for the later hand's pitch-perfect illustrations (fig. 40)? A quiver and a pipe, a snowshoe, maybe a papoose: the story for which Hubbard made these images must have had a Native American theme. Something has been settled, sealed, learned, about, let us say, drawing. But something also lurks, a transparency of a future present tense, even in something so simple as the uncanny resemblance between the arrangements of the images on the page. The letter *M* is a shape with an accompanying picture—a rabbit—whose meaning can't be clear to us without the word that starts the story. The letter *T* makes a tree for a scene whose *O*-shaped moon spherically echoes the first letter of the pictured bird called Owl. Even when we think we're devoting to the merest utility, mood creeps in. It's there, guiding our hand and near-to hand.

Finally, Charles Hubbard's calligraphic daybooks, kept over the course of many years, don't seem to yield much information about the shape and nature of his days. They were expediencies for training his hand to master the art of calligraphy, not least so that he could advertise, "hand lettering has an individuality and a beauty not found in type/your favorite poem may be so lettered." To write the day calligraphically is to hone the day so painstakingly you can only afford a few words. Perhaps meditation as an effect of such repetition makes the day seem to stretch rather than contract, as when it swallows us until we don't know where time went. "August 7. Painted Cloudy. August 8. Painted Cloudy. August 9. Painted Cloudy. August 10. Painted Cloudy." In typescript there's no difference in these accounts; only when we draw the day in lettering is it a difference

(39. *a finger, a yawn, a lock of hair*)

etched in the choice for looping the *y*, up or down, with a horizontal stroke or a curl (fig. 41).

The weather tells us what to record, and the personality inflects the weather in turn. Sometimes the verb that turns Hubbard's days is "taught" rather than "painted," and it occurs with as much frequency.

(40. *the incomprehension of a letter inside a word*)

Sometimes the repetitions are barred like the feathers of an owl with the sudden light of found poetry, as in, "Called on Harry C. on Goose Lane and the moon was round," or "waiting for colors," followed a few days later by, "began to use color on scenes of which there are four."

Fri. July 27 Attended Kumbax . rain.
Sat. July 28 Painted and attended the
Kumbax exercises . Ate beans, rain.
Sun. July 29 . Church p.m. & even. fair.
Mon. July 30. Painted . Mr. G. W . and
two boys had supper with us . fair.
Tue. July 31. Waiting for colors. Called
on Dr. Derbyshire . Drew in line. fair.
Wed. August 1 . Began to use color on
scenes, of which there are four. fair.
Thu. August 2 . Painted . fair.

Fri. August 3 . Painted . Showers.
Sat. August 4 . Painted . Thunder.
Sun. August 5 . Read . had dinner
with Mr. G. W. in the Pines. Show-ers.
Mon. August 6 Painted . Cloudy
Tue. August 7. Painted . Cloudy
Wed. August 8 . Painted . Cloudy
Thu. August 9 . Painted , Cloudy.
Fri. August 10 ; Painted . Cloudy
Sat. August 11 . Painted . Shower. fair.

(41. *a daybook for writing light*)

I've been in the archive so long that I feel my back turned against anything that could come crashing in, hopeful to let unfamiliar moods act upon my world as influence, when I'm introduced to an interested presence at my shoulder. It's the town's historian come to greet and answer any questions I might have. I show him the pages

of Hubbard's childhood sketchbook I've been poring over, and ask if
he recognizes the ledger's names. None are of Guilford born, he explains, able to tell that with a scan of his eyes in an instant. His kinship with the town's details is impressive, but it's the way he reads
the book with his hands that moves me. He touches the spine of the
book with an investigatory caress that cares enough to want to know
it. I'd swear he's weighing it with his hand, rather than exerting his

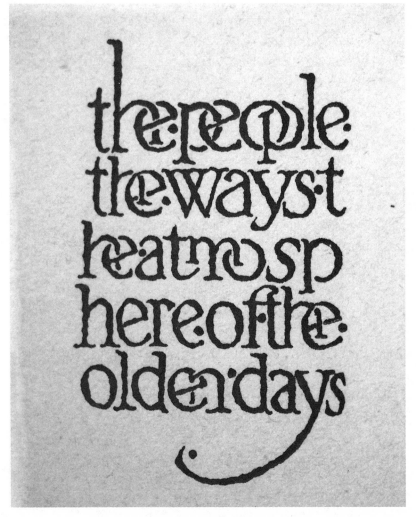

(42. *it was a mutually entrusted mood*)

own weight upon it; his hand asks first what type of book this is; comparatively speaking, what does it wish for a reader to bear? Then he listens the way a doctor palpates and percusses, scanning the pages near to his face, then far, moving across them with his index and middle fingers. How differently he feels and sees than I. There's so much he can give that I, hunched and earnest, fail to glimpse.

Can we hear someone else thinking? Can we hear someone else reading? Perhaps if we could, we'd be able to enter another person's mood, or at least to sense its shape, narrow or open, crowded or clear. We'd have to begin somewhere. We'd have to be glad to try, to wonder about the nature of the rooms they wander in, to discover if they're in any way at all like ours (fig. 42).

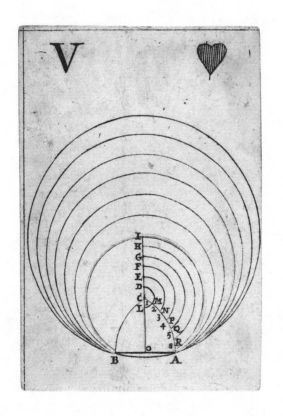

(*vibes*)

Sound has no existence, shape or form, it must be made new all the time, it slumbers until it is awaken[ed], and after it ceases its place of being it is unknown. —Architect Rudolf Markgraf, 1911, quoted in Emily Thompson, *The Soundscape of Modernity*

All that we read and hear covers us like a layer, surrounds and envelops us like a medium: the logosphere. This logosphere is given to us by our period, our class, our métier . . . to displace what is given can only be the result of shock; we must shake up the balanced mass of words, pierce the layer, disturb the linked order of the sentences, break the structures of language. . . . —Roland Barthes, "Brecht and Discourse"

mood telephony

Two phrases greet me like mood budlets upon reading Jean Luc-Nancy's *Listening*: the "auditory apparatus" and the "phonatory apparatus," the means by which we listen or hear and the means by which we produce audible sounds, the to-be-heard, respectively. Phonation and audition create a perfect cavern for Being held like a hollowed-out walnut in the mouth, one word emanating from the back of the throat ("audition"), and the other asking that the lips close gently over the teeth ("phonation"). In this way, the inside of the mouth provides a roof and a room for making meaning, but also for a sensory quietude emitted by the body and returning to it like a hollowed-out lozenge, a pillow-shaped puffball (preferably powdery and peppermint flavored) that is all mood.

Phonation is such a pleasing word for the act of making words, relative to "foam" without the "styro-," a stylus dipped in bubbling eddies, returning mood-forms to telephonics and phone booths that date us, down to the furtive whisper of a mouthpiece. In the period before cellphones, mood was more manifest inside the wires that held the voice and its cradle. Static and crackles were possible, dial's tone more meaningful for its monotone: the ultimate preludic somnambulist without which you cannot call out and only after which you may initiate your insertion into a period of waiting between the time of your rotary dialing and the lift of the receiver on the other end.

Cellphones, for now, in canceling busy signals, in deleting dial tones and other armatures of permissibility and lag, replace mood telephony with feelings: anger, exasperation, punch, and strain. The rotary was the sound of metal on metal impelled by a hand and not its simulacra. Now sounds are assigned to material encounters with a body like decorative labels. We yell, we hem, "you're breaking up!"

It is no longer possible to whisper.

the tic-tic-tic of a dime
hitting the floor

I'm sitting in an examining room while an otolaryngologist reads my chart. I've just turned fifty, but he remarks, "You have the hearing of an eighty-year-old. There's no disease entity that has caused this," he explains, and we've ruled out my having ever been a member of a heavy metal band. "It's more like a natural deterioration that we see in people as they age, but in your case, it's happening much more rapidly." And then he tries to exemplify my future with the mention of hearing aids—which seem acceptable to me—and lip-reading, which does not. There'd been an episode that precipitated my visit to the doctor, but I had waited at least a year or two to make the appointment. It was as simple a discovery as realizing that Jean could hear the pitch of a digital thermometer that was hanging out of my mouth and I could not. I was in the middle of treatment for breast cancer at the time, and this discovery was rather low down on the list of things I was learning about myself. But then I started to notice that I was having trouble hearing my students—which got richer especially when I had to strain to hear them in classrooms where I was teaching courses on literature and sound—until it dawned on me: in the name of "chemo brain," doctors were beginning to admit that cancer and its treatments could affect a person's mood. But no one had mentioned that, if chemotherapy had caused most of the hair on my body to fall off, mightn't it also have destroyed the hairs that governed hearing lodged in my inner ear? Clinicians are only just now beginning to acknowledge exactly that.

When a dime fell from a counter in a restaurant and I did not hear it hit the floor, Jean asked, with a touch of Isaac Newton meets Dr. Caligari in her voice, "You really didn't hear that? Let me do it again." The diminution of a sense doesn't need to be experienced as a subtraction but as an opportunity for an entirely different relationship to the world. (Sure.) I may not have heard the dime hit the floor, but I was ultra-aware of the scent of clam chowder and the softness of Jean's overcoat to say nothing of the intimacy of an experimental mode that forever sparks the mood of our relationship. "I'm not hearing anything," I said as I watched the dime morph from metal washer to plastic bingo chip to cardboard disc, or so it seems retrospectively. When the bounce of its meeting linoleum or concrete—I can't remember which—produced no accompanying sound, I felt frightened but remained curious about the broken correspondences. Dropping the dime a third time and really concentrating, I thought I did hear an infinitesimally faint "tic"—not quite the slap and ting I was expecting—it was more like the way a person with superhuman audition might claim to hear the tip of a pencil tapping through paper on the wooden surface of a desk from a very far distance in an amphitheater. I swear the movement of the object, shorn of sound, was both sped up and slowed down.

Do we ever perceive things as they are?

A dime hitting a floor sans sound, to a hearing person, is oneiric, a word I learned and never forgot from a professor of art history whose hair was as fine and voice as cavernously wispy as the meaning of the term: "oneiric"—like a dream. Some paintings were part of an oneiric tradition, he explained. A painting could change your life, he said, describing one such metamorphic meeting in a museum, but, better, it could fill you with the Rilkean imperative in the poem that he read to us that *you* must change your life. I can still hear him across a gulf of thirty years making lessons in the form of lace at a Vermeer casement. There were bubbles in his voice like beads forever sunk inside a glass pane when he pronounced the words "the ba-roque," prelude to our inauguration into a world of the overmuch. But when he asked us to turn our attention to the "oneiric," he cocked an ear and suggested we remain on the lookout for secular halos that relied on something less immediate than sight. It was mood that he was asking us to attend to, and the only way to get there was by way of his hair and his voice, both of which were bathed in the mildest yellow.

sounding repetitions

"Yeh-yeh, yeh-yeh-yeh, yeh, yeh-yeh."

"The Lower East Side is a fun little area."

"Yeh, yeh-yeh, yeh-yeh-yeh, yeh-yeh."

"Brooklyn is like a bigger Montreal."

"Yeh-yeh-yeh."

"That would be a chill spot, super-chill, cool."

Overhearing puts you outside the zone in a way that isn't always empowering. In the '80s, there was nothing like a Mohawk to make your entire head into an ax, the better to slice through and carve out a zone of differentiation between one generation of people and the next, or one membership and the rest. Today, the jackhammer of the twenty-something's "yeh-yeh" feels more manic to me, less a clean cut to create a mood affinity and more a tic as sign of tenuousness and immanent disarticulation. Nothing discordant can enter this zone of perfect comprehension, the yeh-yeh seems to say. "Yeh-yeh" doesn't mean yes, it means no.

When three-year-old Lea, one of the smartest toddlers I know, punctuates her words with "Um-um-um," she seems to be saying that she knows the adult listener will leave her zone if she introduces too many silences into the space between her thoughts and words. It takes a great deal of stamina to have to leave the mood of her thoughts in order to keep an interlocutor present to her words. As music, the twenty-something's repetitions would ask to be played both *presto* and *agitato*, extremely fast and agitated, but with a feigned smugness and cool. The tempo of Lea's "ums" is also faster than the other words in the sentence in which they appear, but their speed serves as a launchpad for the slowly elongated measure of the rest of her words, *lento*, *largo*. Instead of saying, "I was there," she might say, "I was theeeeeeerrrrrre," holding the note of the final syllable much longer than convention allows.

How much practice is required in order to achieve forms of pitch-perfect self-modulation? And how do our moods get regularized or muddied in the process?

Lea's ums grow faster the nearer she comes to arriving at a real word. The ums are a stall tactic to keep her listener busy with, as I've already suggested, but also to distract herself while some complex inner machinery translates her unmapable mental life into a shareable form. Her ums are anxious insofar as they anticipate her access to a zone where the adults are living, a place whose relative attractiveness is laced for all children with ambivalence. Her ums are tinged with the ecstasy of having a thought, a rationalist's squeal to contain the uncontainability of thought and touched with the pleasure of finding a way to deliver that thought to a potential or particular listener.

Moods must be the products of a process whereby the human subject translates states that have their own time out of mind into the tempos of a common language. Moods, too, a product of the tone with which the ricochet of ums and elongated words is received by another set of ears.

Of course Lea's recurring ums are also her way of performing a form of sheepishness that she's learned is the proper intonation for a girl, especially if this girl wishes to be listened to. None of this is easy. When four-year-old Sam repeats an *n* sound—as in "ne-ne-ne-ne-no, Daddy!"—*allegro*, it's as though he's found the sound to meet the form of a pirouetting "Let me do it!" as when he stands on tiptoe to drop a toilet paper tube into the gerbil aquarium, then time how long it takes for the animals to grind the paper into a nestling atmosphere of downy dust. In the name of gender, Sam will also have to give up the linguistic tippy-toe routine, a forfeiture sure to leave the bad mood of masculinity in its wake, maybe even trading it in for a silent brooding. For now Sam's ne-ne-no's are similar to yet different from Lea's um-um-ums: his is a phonemic repetition where hers is semantic. He's testing a run-up to a word; she's testing the run-up to a sentence. By the time of the twenty somethings' yeh-yeh-yehs, the repetition has come to take the form of some sort of sad bondage, torturing the self and one's listener in turn. The fifty-somethings, though, are worst of all, because maybe the twenty-somethings will create a jagged music from their pain. The further-along-their-course-of-life adults speak a language, properly timed and tuned, in which the animating repetitions of childhood are replaced with the story of their lives and typifying worries. Conversations between adults are endurance tests: how long can I endure the sound of my own voice; how long can I endure the sound of yours? The mood's a solemn one.

the sounds that seals make

The walk took us through verdant musk, or we took ourselves down the gentle tangle of its descent, toe tipping without need of footholds over tree roots melded to rock. It was all inside and growing deeper into the underside of fern coves the way we went until a break led to a trespass through the remains of an orchard, its neglected fruit abuzz with bees, and a sudden bright opening of sky. Across a road and into another wood then—follow it, follow it—the trees more nearly salt-licked, we must have heard a rhythmic movement but paid no heed, or let our burst and jerks of words outpace the lapping: three steps out beyond the trees, a turnstile made of breeze lays bare the surprise of ocean.

The air was clay-y and lichen tinged, the color of gray-green copper; the rocky shore was wet in those places that the sea had recently left, and we were quieted by a red horizon line, one part stimulant and one part drowse, the way some lank paintbrush had marked the lone Adirondack chair red. No one is expected to sit there; what with its trancelike hominess athwart a rugged coast, it might have been washed up; tied down, it bore the signs of placement and care in the making of a scene: it was a mood throne.

The question was how long you could claim it if other travelers happened by, so, in the end, not really deciding, we left it blank—each of its arms a lookout for this bird or that, while each of us partook of a different piece of promontory rock, each to her own, a separate pod of quiet. There were only three of us as I have said, but the chair was enough of a presence to exert the feeling of a fourth. One of us tried to sleep, while one of us tried to read, while the third unfurled a red kerchief, much the same color as the chair, inside of which was tucked neither a sandwich nor a deck of cards but a notepad or sketchbook. She may have been trying to write when the barking began.

Here were no whiskers on dogs or on kittens, no packages tied up with string, but the less common frolic of seals come up from under, splayed on a rock too far to see without binoculars—we had none—but audible enough to unmoor the tonal synchrony that attaches a voice to a human. Even if seals are the opposite of crepuscular (they come out to sun themselves rather than be called by the night), even if they are nothing like moths and foxes lining the darkness with their dust and their stealth, their decision to bring their ink-black bodies into the light and surface remains unpredictable to us, to me, and their voices bare traces of the darker world they've left. Their sounds weren't meant for us, but they cast a tenor-sax-like spell of squeak and squall, making the scene entire into a nightclub part-day, resonant with the acoustics of dream.

A mosquito *was* buzzing all too close to an ear, birds *were* rippling a watery surface with their flap and speed, but we only had ears for the seals. No sooner would they snort than they'd counter the guttural bass note with a trill. A discordant whine would be cut by a gurgle. Then they'd imitate the high-pitched scree of the gull, the hee-haw of horse on land or at sea. Their sounds were liquid and blubbery in harmony with the figure they cut in their underwater world, but their game was to call forth the ragtag babble of everything but themselves: one sounds as though it has swallowed a hinge, while another quacks, and a third one growls.

Casting about for a word for their sound (and with it a mood that she could imitate)—that word was "grackle"—the woman with the notepad woke up though she'd not been sleeping, and the woman who'd been sleeping, drew her arm across the rock as if to write, while the woman who had been trying to read, having forgotten her hat and whipped by the seals' siren song, was red as the empty chair that fronted the rock, ripe to bursting as the apples in the orchard, her entire face now draped in sun-scorched finery. It was decided then that it was time for them to disembark even though they'd never left the shore.

the exciting or opiatic effect of certain words

The hypothalamic nuclei are connected to the cerebral cortex whose functioning underlies *meaning*—but how?—and also to the limbic lobe of the brain stem whose functioning underlies *affects*. At present we don't know *how* this transfer takes place, but clinical experience allows us to think that it does *actually* take place (for instance, one will recall the exciting or sedative, "opiatic," effect of certain words).—Julia Kristeva, *Black Sun*

When Julia Kristeva talks about the sedative or opiatic effect of certain words, I don't think she has in mind the mood-altering capacity of the meanings that accompany words—those bulky overcoats—even though that word ("meaning") appears in her gloss. I suspect she's talking about words as sound forms whose texture and timbre have the power to, as the saying goes, touch something in us, and in the touching, either to create a new mood, if such a thing is possible, or to conjure the residue of a mood that's gone missing. Psycho-phonologists read high-frequency sounds as capable of producing states of heightened awareness in we humans, acting as they do on the cochlea, whereas low-frequency sounds can calm us to the point of stasis and torpor: if the liquid inside the semicircular canals of the ear's vestibule is made to rotate enough, by repeated low-frequency drumming, say, a state of trance is the result. Then we are said to be "captives of our vestibules."

But what about the effect of language on our "neurobiological networks"? Is it possible to identify words that at one time made us happy, exclusive of "candy"? And how about words that exert a drone or din? Just as worry is easier to bear in a particular place, so worry is easier to bear surrounded by particular words. Can words in themselves have this power or does it depend on the quality of the air through which words move? There's the rub: doesn't it all come down to voice, the ineluctable wooing of one by the other—word and voice, ear and tongue and throat, lips and lungs? If pronounced in *her* voice, all words create the best mood in me. That's the ticket. All distinctions fall away.

Use the next full minute to list words that come to mind as likely to produce a soothing or pleasant mood in you. Go!

denizen versus citizen
hula hoop versus tire iron
glockenspiel versus man-o'-war
harmonica versus accordion
charlotte, but now we're back to ice cream, or dessert.

Swarthy, swatch, and glade; recluse and surcease; recant and disuse; delve, shelve, elve; elevate and conjugate. Jugular and jaguar. Constantinople. Fructify. Gina Lollobrigida. Riffraff. Rinky-dink. Edgeless. Leavening. Sausalito. Somersault.

The mood-producing effects of such words must have to do with the nap of each person's individual fur, each person's causeway-like zags, marbleized or plush, the orientation and density of our inner and outer linings. And maybe, too, with the mechanics of an accented rise and fall of the voices that originally coaxed us into being, "Come out, come out," they said, "for now it's time to come out." Or, "sleep, now—there, there—it's time to sleep."

We leave it to poets to return language to its roots in the body, to restore language's place amid the elements, earth, air, fire, and water. A sentence can move as mesmerically as a reversing falls falls like the small and quiet ones hidden inside trees more majestic than those that pound pound for pound and measure for measure weight of their force drawn down down or up the sentence sentenced to reverse itself to meet but not to find itself again drawn back upon itself not by itself alone alone upon a pad this pen and that heart draws it forth and back until a feeling is produced by it and then it stops.

We turn to poets or to the poets we, ourselves, become when called to attention by distillates even in the most analytic prose. Then I gather such phrases for their capacity to say everything that needs to be said, that are in themselves all the mood-thought we need to understand depression (for example), as from Kristeva the words pools,

"institutionalized stupor"
"prisoners of affect"
"the delights of suffering"
"nychthemeral rhythms"
"our most persistent despondencies"
"to tame and cherish sadness as an object for lack of another"
"a lucid counterdepressant"
"to unfold language's resources"
"our basic homeostatic recourses"
"faced with the impossibility of concatenating"
"learned helplessness"
"playing dead"
"psychic crypts" or "psychic voids."

In order to effect a mood out of language, need a writer put words through the same process that herbs are subjected to in the creation of mood-enhancing cordials? Steeping, distilling, infusing, and macerating, all of which share the requirement of soaking and softening, condensing and extracting, supply the idea of a liquid aesthetic, and who wouldn't wish to produce in a fellow being the combination hum and high of cranberries soaked in bourbon?

Maybe a poet's charge is to unsteep words and in doing so to perform an only seemingly simple operation of extraction, to allow us to hear what we never hear inside the words we always hear, for I know I am put in a mood part joyful and part curious—not an opiatic mood but a wakeful one—when met with the word "seemly" over and against the more commonplace unseemly; when prompted to imagine a "shevled" rather than disheveled appearance; to be made to consider what "whelms" me as distinct from what overwhelms me; to comprehend the way in which each repetition is a renewed "petition"; to find loose-leaf pages—a "quire"—at the center of all requirements and inquiries; to posit positively against the force of certain words' tendency to exist only in negation—to eke out the "choate" in the inchoate, the "ane" in the inane—to "bibe" and "bue" without a consuming "im-": to saturate.

Occasionally you'll hear it on the radio, how the confident mood of capitalism turns hysterical. Then mad throngs storm the vestibules of Walmart intent on a wide-screen TV, trampling to death a guard in the process. What words create the frequencies to inspire mass motility numb to the sound of voice and tongue and throat, lips and lungs, heart and mind and memory pulsing underfoot? I think of Thoreau's different drummer, of Dickinson's poetry of tilt and whirl. I wonder if poetry undelivered, distant but there, poetry requiring that we crane just long enough to pause indeterminately can avert the disaster of stampede.

arrangement for voice and interiors

Leave the errands in a basket attached to a bike's handlebars in your imagination (the real bikes were stolen and they never had a basket). Put a little dog in the basket and picture life in a village not far off. Place the poppies in the foreground (it could be your own backyard). Place the chanterelles against a leaf vein broad and round, place the whole upon a bright red surface. Place, don't put, the white puff of a small fall flower against a bramble's stippled branch; let red berries brighten before dark bricks. Here's a wheat frond, blonde, and a brown wood fence.

Some days, I find, ask to be arranged and rearranged, assembled and reassembled especially if I've managed to weave a cocoon to make a sentence in, or a fog to retreat into with a book. Then something that I've not only scheduled but have paid for and therefore can't afford to miss requires that I leave. The problem when this happens is that I find myself carrying bits of pupae into the outside world, my feet stick to the ground as if I'm trailing goo, and the place where I last left off departs. The words of missed pages begin to spell themselves backwards like Scrabble tiles poured back into the bag, or the book closes back to front rather than front to back while I walk forward, or try to.

I'm compelled by still lives when this happens or the compulsion to make such out of the things I encounter along the way as if instead of losing me, the mood I've left will bear itself out in some future just ahead so long as I keep on the lookout for disparate parts or keep an ear open to the chalky baseline of an echo.

Certain houses have opened their doors today for the general public to come in—the stately, sturdy, off-limits sort you wouldn't generally have the opportunity to see inside of if it weren't for the fact of this neighborhood fundraiser. There's only one way of moving inside the habited rooms of strangers who have deigned to let the general public in—who knows exactly why?—and that is by way of creep and stealth, of tamped-down enthusiasm and borrowed awe. Still, you brighten to the beauty of these places, their history, their lure, their preelectric darkness hung like drapes, their narrow stenciled hallways, and steep stairs.

Inside each house tour there are voices—booming, stern, "No pictures allowed!" an organizational rep yells. A voice assaults you at the top of a stair, accusingly, like a cop, "I know *you*!"—or was it "*Don't* I know you?" "I know you, too," I say, but neither of us can recall from when or where. There's a carriage house missing its crop and stir, evacuees of sound. The creak of pulleys muted, the whinny and the shudder of nostrils, hooves and flesh long gone. In one house, the owner himself stands in his own front room. He's surrounded by parallelograms of "eep" and "aw," of mice-like voices pitched too high and sipping cider. His house doubles, triples, and quadruples: there are two kitchens side by side, sundry perennial gardens, there's a white soup tureen atop a white broad plate atop a white table cloth on a white table set on a white glass floor. When I reach to shake his hand, he fails to offer his. It's like a blow to the throat, and I hear myself saying, "Oh!" Just that, and nothing more, audible only to myself: "Oh!" As if I could try to understand the literal refusal of an outstretched hand. What I'd really meant to voice was, "Wow!"

Place the poppies so they can tell the way to the house because they do not rise high up from the ground, they do not climb, but flag and thrip like tissues on a stick, they wave so that fronds can jut and broom-bush bend behind them every which way and the sound of the insects rubbing their parts can make the hot bead at the day's center, heating the house tiles to boiling, baking the tubular roof's clay. Front the ocean's chilly dark with this wave of blackly centered red as signal and chord to a distant hearing, a further sight line.

Adjust, situate, rest, lay, position, join. Turn into this final house before returning. Our obligatory stroll was coming to a close but there was still a house-turned-lawyer's-offices to see. This would have to be the least interesting interior because no one lived in its rooms but only met to churn out paperwork. There were sounds trailing me (sentences from a book); there were sounds up ahead (cellphones); there was a still life: four tumblers meant for bourbon on a tray; two high-backed knobbed leather chairs; taxidermy. There was a mood of glut and emptiness and gout. In place of a liquor cabinet, a liquor vault just visible through a door, a liquor walk-in closet with rows and rows of shelves made of glass; gin's sapphire-tinted bottle; green Tanqueray.

Positioned inside a sea of *Ice Storm* family photographs that were strewn about the room, an aged man.

He had the look of someone whose faculties were held on by a thread so he tottered even as he continued to reach, and before I knew it he had pulled me toward him with one hand in my hand, and another on my wrist, he explained he was the lawyer's father—the lawyer was gone for the day—and there was something he wanted me to see. I looked back hastily at the other people in the room—wasn't he supposed to be giving a tour?—but it was as though he only saw what he could draw into a zone fast before him and I happened to have trespassed in.

He drew me to a photograph. We both stepped back a little, then leaned in. He spoke in a style that had fallen into disuse, not that it had ever generally been acknowledged as a style. It was a pattern I recognized from my place as a child at an Italian American table: just before eating or after the meal, someone produces a disquisition beginning with the gravy. "See this gravy here?" it starts. It might have to do with the origins of tomatoes—not historically, but in the garden behind the house—it begins with tomatoes, beautiful and plump, but it ends with a pain in the side, deepening to a graver suffering even though the storyteller smiles while he tells of it. What was left of his voice was a gravelly, phlegmy whisper, but it still could muster such mixed tones—of enthusiasm, apology, and grief.

The photograph was nondescript to me, but he was asking me to see the world the way he saw it. It must have been why he'd agreed to show his son's rooms, or perhaps in spite of his own best effort not to go over the story again, he sensed a sympathetic vibration in brown eyes. No, he sensed my own precariousness, my search for a center from inside the disarray.

The photo was a generic group shot in black and white of six men arrayed in two rows. One of the men—but they all looked alike—was supposed to be recognizable as the towering figure of his father. Strapping he'd say, strong as a bull, indomitable. He wanted me to know that his father had been the fiercest handball player in the world. He was unbeatable. He broke the ball.

Nothing could touch this man. He had achieved acclaim. He wanted me to know this. His father died young, that was the crux of the story, and he, the son, was never able to make sense of it. His father had contracted an infection in his heart, and the Providence doctors said there was nothing they could do to help him. But the son didn't believe this so he drove in his 1978 Chevy Malibu to Boston. Distracted by his father's condition, he hadn't been paying attention to the news, and he barely made his way to Boston in a blizzard that would go down in local history. The city was a ghost town. Even the hospitals were closed, and by the time he saw the doctor, his father had already died. His father had died, but he'd carried his X-rays to show these better experts who told him they could definitely have saved his father and that he hadn't received the proper care.

He could not hear. He hadn't let go of my hand. Many people entered and exited the room, peeked in to the liquor vault, and left. "He shouldn't have died," he said, "nothing could kill him." "I know," I said, "I hear what you're saying. I can see he was a very strong man." "I know," I could've said, "I hear what you're saying, but you couldn't have saved an indomitable man."

Put the Chevy Malibu inside the poppies and hang gray trunks on

a stark white line. Stow a stranger's sorrow inside your mood plan—
but you forgot that long ago. Now there's only the voice of disarray
and failed arrangement: would his father have lived if he'd arranged
the day differently, if the weather had agreed to line up with his
plans, if only he'd made other arrangements, if his voice could have
reached some savior through the broad, white bands of storm.

sonorous envelopes

At five weeks, the baby can distinguish its mother's voice from others, though it still cannot differentiate between its mother's face and those of others. Thus, before the end of the first month, the infant is beginning to be able to decode the expressive value of the adults' acoustic interventions. —Didier Anzieu, *The Skin Ego*

At this moment, you are reading. I am absent. Still I shall pretend to talk. —William Gass, "On Talking to Oneself"

"Let the matter work on me to create the total scheme": whom do I talk to when I talk to myself while driving the back and winding roads of Slocum, RI? I know the route well—it's my way to work, which has always also been a way to school, to teaching and to learning. I'm to be present on this day to a ceremony, and I'm distracted, knowing as I do how far obligatory speeches and pomp really are from the intense intimacies of study and discussion that happen in the classroom, in so many cases giving way to a feeling felt for the first time. Writing can never be continuous with life but a separate road that rides along life's surface, so it helps to be on a road I know fairly well if the thoughts that can lead to writing are to drift in through some window. I have to get somewhere and by a particular time, but I'm thinking, "Let the matter work on me to create the total scheme rather than map the book and, so, control it."

I'm pitched to absence, to openness, to aloneness when life breaks in like a gift from the gods: it's the middle of May, the roads are scented with breeze. The winter had been particularly harsh, with plummeting temperatures and blinding drifts as if in anticipation of the gentler showering of this *now*: white puffs float past the windshield in droves; unpropelled but carried, the entire scene's astir with them. It's snowing neither petals white and pink nor dandelion seedpods, those childhood wishes in the wind; these are puffs from cottonwood trees, or so I decide as I drift toward what I think I know from books. Floating, they transform the road I know so well into a greeting, a cloaking, the obverse of a beckoning salute that rims the edge of Elysian Fields. Here there is no river of forgetfulness to cross; today I'm traveling a road meant for remembering the oftenness of the route and of the journey—try twenty-four consecutive years, save for a break in medical crisis, in mourning, in turning away to take a different path once every six years. How often the atmosphere here is rowed with maple trees; jets that pulse to wet the turf fields; always there is carrion, some small animal felled by obtuse cars. A lone copper beech has thrived.

I know there is a tiny rippling falls folded into the underbrush, a pond hidden by a house, and it's at that bend, with or without the whoosh of a hurtling train whose trail runs parallel to this route, that I feel it: a nervous wrench in the gut, and the need for a bathroom as prelude to some performance just up ahead and beyond the parking lot. Often there is a tailgater who isn't seeing what I am seeing—or it could be myself, that time that Jean wanted to stop to photograph the jack-o'-lanterns, dozens of flickering gap-toothed faces atop the fault lines of Puritan-built stone walls, when I just wanted to get home—typically there is the impatience of some other to be dealt with or ignored, especially today in this floating fifteen-mile-per-hour mode, I must seem senile to fellow drivers when I pause to open the car window to the thought: "Let the matter work on me to create the total scheme."

So many favorite writers had already set the mood toward which my pages tend, and maybe most especially Roland Barthes, the paragraph happened upon just when I needed the words to say it, exclaiming what I'd been after in the first place: a route with no destination except for the ability to sustain and withstand a mood of *being alive*: could anything be more obvious and less evident, as when Barthes describes it in a journal dated July 16, 1977: "Again, after overcast days, a fine morning: luster and subtlety of the atmosphere: a cool, luminous silk. This blank moment (no meaning) produces the plenitude of an evidence: that it is worthwhile being alive. The morning errands (to the grocer, the baker, while the village is still almost deserted) are something I wouldn't miss for anything in the world."

As for this feeling of being alive, there was no moment in my childhood when my mother wasn't holding my hand. She'd be holding my hand and lost in her own thought, drift-ward, up ahead and out of sight, or backward-turning to what had left a wound or couldn't be recaptured. She was like a kite, glorious and shape-shifting, to which I was directed to hold fast. If I let go, I might experience my own lift off; if I let go, I might tumble head over heel over head endlessly to the ground. Her voice was billowing and blue.

The atmospheres created by the voices of our earliest caretakers, vacant or urgent with calling, heightened by emotion, piercing, riotous or fraught, pleasantly clucking, caught off guard by their own surprise, fastened by a thread to calm, impulsive, doting, and alarmed: the envelope of meaningful sound that both holds and addresses us from the start is mood's point of origin and of no return. To arrive at it thematically is to depart; it is a point of encounter and a line of flight.

If I fly with all the might of a baby bird in search of a landing point, I arrive where I started: with a childhood interest in deafness. By eight years old, I had developed a fascination with a sealed-off world or one only accessed by vibrations; I was intrigued by the idea of a different sign system than the one I'd been taught. I fancied finding people who could teach me their language—the language of those who could not hear—and how they'd welcome me into their community. The rooms, as I imagined them, would be small and dark, somewhere in the vicinity of Washington, DC. Though I had not made a conscious connection with the lure of cryptographic messages and spies, I imagined our communicating via a means apart: using only our hands and radio waves.

My situation was this: if my father, the perpetual yeller, had made me deaf to my own voice, my mother had made me into a listener of the highest order for the way her voice called and delighted, but it had no time or respect for silence, and sometimes, though it appeared to address me, it didn't require my presence: it did not know I was there. Since then, I've come to understand a certain kind of voice modality that must exert itself full-on in order to create an atmosphere around the speaker long before any real communication can take place. Whether an interlocutor will wait around for the establishment of such conditions—all the while listening or pretending to—is anybody's guess, but there will always be at least one person whose station in life is to have no choice in the matter. If you think what I'm describing is the situation of someone loving the sound of their own voice, think again: no one does, and people who talk to excess, least of all. Such people, I believe, are awaiting the sound of their voice; they are unable to hear it or are unconvinced of its existence.

My mother writes the letter that I dream. If my mother, a poet, was writing, everything was OK. Years ago, I'd written an essay driven by that truth, set in the era during which my mother, suffering from agoraphobia, used letter writing as a means of leaving the house. I'd know by the sound of the gas stove ticking, and the smell of late-night coffee brewing, that she was in her element, that terror was temporarily at bay, that I could sleep. That was one order of voice I could always listen out for to create a soothing mood: the sounds that were the signal that my mother might be writing. Otherwise the most powerful mood cauls were those cast when I was sick, separated from but within earshot of the voices of adults.

Everyone must have an early childhood experience that puts them at a far distance from the shore: when I fell from the top step of a high diving board to cement at seven, my hearing was temporarily affected. Blood had trickled out through my ear making the voices in an outer room seem swollen and muffled. The residue of those near-far voices, rising and falling, created a net or mesh made of all the tones that would necessitate my own future writing. Unlike kids who would beg to experience the privilege of staying up with the adults, ever since then, I preferred to listen to their voices from the vantage of my bedroom. I don't want or need to decipher the strange language that they speak; nor do I long to be let in on its secrets. I keen to it as a kind of night music or familial jazz laced with pleasure and sorrow. I make of it what I will in the where of my own becoming.

It can't be helped. We might grow up to alter or create soundscapes—shareable new musics that we can call our own—but the impetus for such creations is to be found in the human voice. The layers mount like filo dough, for, even if each person's voice appears to be her personality's most vividly differentiating distinguishing feature, the poetry, art, or music we come to make is not born of a singularity of voice traceable to the self. My voice—the thing that makes me, me—must be an amalgam of numerous voices I have heard and internalized; an imitation of those voices, the reply to an intimate listening that had love and survival at its center. My speaking voice is radically my own, and radically other, and it is different and the same as the voice that I think with. My writing voice is something different yet again—it's the stage upon which a different voice is loosed: it's the condition of a possibility for song, whether my physical voice is capable of holding a tune or not; it's a voice as interested in listening as it is in achieving a voluble note, a voice turned in the direction of the silences and the murmurs of voices in the outer room of those people who had cared for me, or held me, or lived their own unarticulated longing, nearby. It's a voice forgetful of itself and hopeful for something it has yet to hear, or say.

Bad writing might be nothing more than the voice-as-wall I build around myself to prevent unwelcome sound. All along, I've been pursuing the idea that mood is equivalent to how voices hold us— think of how especially apparent this is when we're both stirred to hopefulness by a voice from the pulpit, moved to act and cradled all the same, or when we can't get enough of the voice of a writer at a public reading—"I could have listened to you all afternoon": one type of reading turns our heads in holding us while another turns us off or puts us off, repelling us. Forget phone sex: I can experience an orgasm based on the *sound* of a particular voice through the phone, the content of the conversation is negligible. Now I wonder if mood is something that antedates such states: mood as inextricable from the voice, not merely its by-product. I'm thinking, again, of the way that earliest voices form the ground of our felt-ness and embodiment or skin, the "sonorous envelope" requisite and precarious, the carapace that is our mood.

I know it is possible to train oneself to listen differently—we can do so as ideological subjects, as maturing parents, as musicians in training, as gurus newly focused on the sound of silence, or as aging bodies bent on overcoming lack. And we can train ourselves to speak differently—less volubly, less often, with or without a regional accent. Strapped inside an MRI machine, face down, I forget how harsh the hammering is and how easy it is for sound to create a feeling of walled-in helplessness. I can shuck off the imperative of this particular mood room's décor by proving I can hear myself think if I concentrate; I can breathe without having to hear myself breathe; I can suspend the need to attach a sound to a thing to make it less threatening—this sound is its own thing. But I don't know if it is possible for any of us to alter our own sonorous envelopes as blueprint for a repertoire of moods.

It's a bit like the impossibility of being able to hear one's own voice, the impossibility of singing oneself to sleep, or crying oneself to sleep, though, according to a great deal of student poetry, the latter is possible and it happens frequently in dorm rooms across the land, mostly of women. Crying ourselves to sleep, we become babies all over again, soaked with sweat upon a cold wet pillow, exhausting ourselves with waiting for the figure who never comes to our call, hiccoughing loudly, and gasping, into the regressed silence of a weightless repose.

Rather than cry myself to sleep, I take solace in the company of happy philosophers—the sentences of Peter Sloterdijk, for example, the tone of whose work I find unrelentingly confident with joy, or inspired psychoanalysts who practice as much as they write, for instance, Édith Lecourt. Without quite submitting to the blame-the-mother routine, Sloterdijk projects a maternal imago we can all aspire to even if we are doomed to fail. The mother must create a "sounding pantomime of happiness" in welcoming the child into the world, feeling that "she must be happy for the child's sake": "Act in such a way that your own mood could at all times be a reasonable standard for a shared life: that is the categorical imperative of the mother." Voice oneself this way—comport oneself—not only after the child is born but for the full term that he is in utero as well. Sloterdijk goes on, "The child's state as the object of the mother's expectations is conveyed by audio-vocal means to the fetal ear, which, upon hearing the greeting sound, unlocks itself completely and takes up the sonorous invitation."

Édith Lecourt, in "The Musical Envelope," gives us the felicitous image of the baby who "gives and takes his note amidst a sharing of sounds (noises, music, words), vibrations and silences," and she leaves us with a poignant distinction that I consider at the heart of any voice-mood-skin creation, elaboration, or experience: she discusses the creation of a common zone of the sonorous bath or sonorous envelope (she uses the words interchangeably) created by the mother's voice, which, though originally affiliated with contact, "will subsequently exist on its own, without the body contact that accompanies it, becoming a communication over distance and even in absence." A human voice cannot come into being without this sonorous feedback, but what seems more redolent with the haunting of mood is that necessary gulf that opens between the point of contact and the point of absence and departure. If vocalization begins as coterminous with contact, it ends with a ghosted memory of touch-as-voice.

It is as plain as day: none of our mothers—or in lieu of mothers, our earliest caretakers—ever really dies, for their voices are imprinted on our skin, not just in the ways we come to sound. In this way, our mothers are ever with us. Is it possible that, unconvinced of

this, unable to know what it could mean, we write the mood space
that is a lifeline to the voice perched on some far off promontory,
or absent altogether except in our ability to conjure it with our own
voice, our own skin?

Working with a mildly autistic child who communicates with
her sublingually and via the piano, Édith Lecourt describes a process
whereby she finds ways to "hold" the child without literally holding
him, eventually inviting "to support him from a distance and watch."
A prototype for the proverbial "letting go," she is preparing him for
the absence of her voice and of her gaze: he can hold himself up,
she seems to say, as all human caretakers must encourage. He can
wander around inside his own mood space and create his own mood
tones in turn. Would that it were this easy, but she knows it's not.

"If the mouth provides the first experience, brief and vivid, of a
differentiating contact, of a place of passage and an incorporation,
repletion," writes Didier Anzieu, "brings the infant the more diffuse
and more durable experience of a central mass, a fullness, a centre
of gravity." In such moments of satiation, a baby—and later adult—
might feel confidently moored, but this doesn't preclude a desire to
be transported as with music, an experience that Lecourt links to
the weightlessness of being carried in infancy. Coming into a skin
that makes possible a voice and, between them, a mood, we seem
bound to teeter between a need for ground and a desire for flight—
we're forever caught between gravity and gravitas, flight and floor.

~~~~~~~~

I once played, for my students in a literary acoustics class, audio files
made by astrophysicists who were attempting to simulate the sound
of the big bang, that event whose effect was the formation of the stars
and galaxies. Without telling them what the sound was supposed
to represent, I asked them what they thought they were hearing.
Though none of us had ever heard a bomb dropping or the engine
of a war plane—in itself a register of our insulated privilege—we
thought this was what we were listening to—that, combined with

the hiss of a snare drum and the ever-diminishing sound of a coin dropped into the gyre of a spiraling well.

That listening and the discussion that followed could have yielded volumes, but what was most daunting in its import was the question of the what and why of the physicists' experiment, since, for one thing, it wasn't clear to we nonphysicists what exactly was being simulated: The sound *at* the beginning of the universe or the sound *of* the beginning of the universe? The sound before sound? The sound at the beginning of time? The sound of creation? The sound of a theory of a universe's expansion? The sound of a sound beyond human comprehension? Taking a meta-approach to the experiment, we asked ourselves what might be motivating the impulse to simulate the sound of the big bang in the first place.

The physicists might answer that they believed they could derive the structure and properties of the universe from their fabrication of or intuitions about an originating sound. From a psychosocial point of view, though, we could say the experiment was indicative of their desire to be somewhere where they could not be, a desire to be present at a point prior to existence as such, a time outside of time. They were using sound to mediate their own relationship to a fundamental absent-presence, to nothingness, and to inaudibility.

The questions that compel us about mood, gravity, and sound depend on where we're standing, and everyone will recall the ancients' beautiful preoccupations with trying to conceptualize a music of the spheres that could map itself onto humors in the body, cosmic laws that could align themselves with bodily and spiritual fluxes and flows. The poetry of these theories is enticing as when human beings are understood to contain the qualities of the stars, or the universal life force is found in plants and nuts: "Nutmeg contains the qualities of the sun's rays, peppermint the combined qualities of the sun and of Jupiter."

For centuries, moods were understood to be the effects of our elemental relationship, grounded and awestruck, to bodies of light and dark that moved with us and athwart us, that governed our fluency or torpor, that made us quake and made us sing. In the twenty-

first century, if contemporary art films can be relied upon for evidence, we are presented with a radically different optic: now the
universe, outer space, intergalactica isn't a zone in relation to us but
a place we wish to escape to in order to have feelings that we cannot have on earth. I come to feel as though in the hands of some of
the finest contemporary film artists—can they represent a collective contemporary desperation?—we are asked to catapult mood to
the stratosphere because there is no space in which to experience
mood on our planet. Projecting mood into the soundless firmament
of outer space, does contemporary film-art fantasize a mood set free
from its dependence on a sonorous envelope? Or is the projection
of mood to a place of radical silence a rehearsal for finding ways to
experience mood in the current white-noise free fall that is life in
the Internet's age?

There is Lars von Trier's visually lush *Melancholia* (2011), whose
characters literally slog through the swamp of nonfeeling that has
become the earth until the planet is altogether vanquished and all attempts at what used to go by "emotion" return to a primordial mood-
dust, the aftereffect of a contemporary big bang. There is Terence
Malick's interminable *The Tree of Life* (2011), in which waxen, mute
families perpetually bend backwards in the hope of being delivered
to some place of choked sadness out of view and out of earshot, far
beyond earth's skies. There is the beautiful fantasia of Alfonso Cua-
rón's *Gravity* (2013), which, though it ends with the slow movement
of feet through sand, like a baby's first steps in some soft and watery
terrain on an uninhabited edge of a planet unrecognizable as earth,
suspends its only character in a gravity-less, infinite expanse for the
full course of the film so that she can work out feelings unavailable
back home. Only in Patricio Guzman's gorgeously calibrated *Nostalgia for the Light* (*Nostalgia de la Luz* [2010]) is a political dimension
woven into a contemporary preoccupation with the heavens when
he juxtaposes two seemingly incompatible quests in the Atacama
Desert: those dedicated to studying the skies who look upward for
answers, and those who look down into the dunes where, for de-

cades, they sift in search of granules of their loved ones who disappeared during the reign of Pinochet.

In that film, mood is not projected symptomatically into a world elsewhere, even though the desert and the skies are literally and figuratively templates for quests without end, for dread, for loss, and for hope. It's as though Guzman has gotten at the philosophical and ethical heart of all of the other galactically imbued art films by documenting in tandem those people driven and beset by an impossible mourning and those people driven and beset by impossible questions. When the mood of scientific curiosity and wonder—laced with an animating or stultifying uncertainty—meets the mood of a galvanizing grief, an ethical imperative is born to work against forces of erasure—of history, of memory, and of time.

Is our yen for an elsewhere something new? Is the desire for a feeling state that is unavailable here, now, a problem more acutely experienced in the displacing no-zones of the information age? I'm not sure. My own compulsions are scenic: just now, a desire to return to the frames that close a film from a much earlier era in which a small boy in a very large room turns away from his mother—she has been too unreliable, she has caused too much pain—toward a vanishing point at the far end of which may or may not reside another human figure, or at least the mirage of one (in this case a new friend who is an older brother, teacher or monk). In Vittorio De Sica's 1944 earthbound *The Children Are Watching Us* (*I Bambani ci Guardano*), the adults are tortured aliens whose incomprehensible demands stretch children's capacities for safety and for love every which way. Here the problem of the sonorous envelope is that it keeps getting torn the moment it is established: the (mood-making) world that the child is welcomed into is at the same time the world the child is ousted from.

What word could do justice to the feeling conveyed by the face of the boy when—wordless—he turns round and walks away from his mother and into the great expanse? The way De Sica shot this scene, the boy could be walking into a Renaissance painting at the moment that inaugurates perspective; he could be entering the draft of a high-ceilinged mill or warehouse manufactured by the industrial age. The vast hall seems even vaster as he turns, but it holds him like a dot of light in a wildly unvariegated field of gray. He tilts his head to one side and turns it down to face his heart, as though a string has loosed and he might never be reanimated. His lips are tightly shut and he seems to breathe through his shoulders and his drooping hands. The corners of his mouth collapse into something other than a grimace or a pout. It could be the sign of a quivering bite meant to keep himself from crying. It anticipates some future rotundity and the cloaking moustache of a man. Turning only his head, at first, away, he neither runs nor walks confidently. He moves slowly, as deliberately as any child is able, uncertain whether his footfalls in that large hall will barely be heard, or not at all.

~~~~~~~~~~

So many *en-* and *em-* words are about being held just a tad too close—"embroil"; "encase"; "entrap." Entangled with another in sound and voice, we fail to knit—we are knotted and twined. Enmeshed, we're stuck together, screen sunk into screen. Embraced departs a bit: we are fleetingly supported and affirmed—we are kissed. Ensconced, we are steeped, but what I hear inside the sound-form "envelope" is vellum and velour, volume and voluptuousness, elopement: with the image of the sonorous envelope, the voice that originally calls us by our name is a papery pouch, and we are the letter contained therein.

Describing the retrograde retardation of depression, Julia Kristeva paints a picture of a subject cut off by dead ends and prodded with unavoidable shocks inside a labyrinth: "When all escape routes are blocked, animals as well as men learn to withdraw rather than flee or fight." Then I imagine the sonorous envelope as engulfing and entrapping. "Tricyclic antidepressants," Kristeva notes, "apparently restore the ability to flee," but what about the ability to restore or remake the sonorous envelope?

My clinically depressed brother habituates his days and any

people willing to listen to him to a daily recitation of hypochondriacal bodily symptoms. In thirty years, there's hardly been a rend in the need to recount himself this way, and he's spent much of his life (he's now approaching sixty) threatening to live or threatening to die. To witness the disease is to stand helplessly by while the loved one undevelops rather than grows, but even that's not accurate because we're all stuck in our ways, bound to the habits that secure our personalities, and are rarely changeable. In depression, something in the process of winding up becomes unwound; the careful creation of a molding becomes undone, and in their stead, lives a kind of intractability as life's work that gives a person license to retreat as far as possible from the world. That's at least the lesson of depression my brother has narrated to me most of the decades of his life. No drugs can really reach his vertigo—and his vertigo is permanent—because, as I've come to know, the world does not stop spinning, it does not pause to wait for any of us to catch up to a feeling. A search for a shadow—the ghost-as-cause—is futile, like the stalking of mood itself, but mood morphed into a bad or evil twin. I wonder how, in such cases as my brother's nearly lifelong suffering, the voice as a *conditioning*, adjectival, flow morphs into an inescapable *condition*, a nominating thing.

My brother wasn't always this way. There was a time in the early seventies in his midteens when he taught himself to play Steve Howe's "Mood for a Day." He'd stand on his toes and lean a little closer to the guitar—that sounding board—right before playing the harmonics—those angelic chords—of Yes's "Roundabout," which he'd also taught himself to play, but the instrumental piece "Mood for a Day" was a point of pride for him, the way he came to it by himself, and mastering it, reembodied it in solo performance. For days our house rung with the intricate, wind-borne tune.

The keynote to "Mood for a Day" was "just listen," for that's what the untrained Steve Howe had done. Though my brother could read music, Steve Howe could not and claimed only to "use his ears" in the composition of what came to be an iconic song for the era. So my brother listened, playing and replaying the Steve Howe LP until he could locate the notes on the neck of his guitar, stroke the strings—glissando style—with all five fingers, percuss the body of the instrument as in flamenco, pause to create the quiver of a tremolo, suspend the right hand to sound the basal G string with nothing but the thumb of the left hand.

This was an era when men could shake out their long hair—each man a slender stalk in his sequined cape or poncho, his long and delicate fingers lost inside belled sleeves. The manual dexterity required by "Mood for a Day" was breathtaking, and Howe's renditions of his own song allowed for a great deal of slippage. It's not a song that asks a performer to play it cleanly, nor to aspire to the reserve or precision of the perfectly plucked note. It gives the feeling, instead, of someone adept at self-accompaniment. It wouldn't mind if you happened to hear it coming through a window—in fact, it may have been intended to be listened to this way: like overhearing someone caught up in the private celebration of a beautiful day who has decided that the best response to such a day is to practice the guitar. Add to this the affirmative relay of the band's various names—they considered calling themselves Life, then World, before arriving at the blissful delivery of a Yes, and you have a kind of recipe for mood light.

My brother did give himself this gift once: he played the song as if literally to knock on the door of the guitar and invite a feeling self in; he found in the guitar a resonant body, perhaps his own. It was the same era in which he passed, with our other brother (ten months older than him), through a weightlifting stage, and a Harley Davidson stage, and a drug-taking stage, none of which achieved the intended outcome of making them anything other than bespectacled sissies whose voices never stopped cracking. They were wimps, god bless them; they were geeks, and well do I remember the gender bend of my slack-shouldered, long-haired brother at home for a spell inside his "Mood for a Day." Our father would have flung the garage door open to greet the change of season and the windows would be sprung to let the familiar sign of a neighbor's braised meatballs and crushed parsley in. Our father would have appeared in the yard slunk close to the ground at work on the patient process of unbending whatever winter weather had trampled underfoot. Indoors our mother would be lost inside the clutter of her notebooks at a desk in a corner of the dining room. How often had my brothers and I gathered at that table not only to eat, but also to perform our homework in sync, only pausing to rotate the arm of the pencil sharpener my father had mounted to the wall.

This is how the memory of my brother's playing makes the rooms *feel*, one sonorous envelope begetting another: my father measures out seeds on a tin table in his garage like a pharmacist and the tic-tic-tic of something vegetal meeting something manmade measures out a beat in time with the ball bearings that shift inside my roller skates when I walk shakily with three stump steps, then roll down the driveway unnoticed but nearby. So my brother for a spell created mood rooms for us, for me, with his guitar—this tripping down the lane made possible by the mood created by his guitar, tethered, untethered, faltering and free: he was playing it for himself, I hope, but it set a mood for me.

~~~~~~~~~~

If I were to assemble a still life that could create an image repertoire for mood and voice, it might consist of three tableaux: the bedside scene of Philip V of Spain whose melancholy was famously relieved by his being sung the same arias each night for fourteen years by the castrato Farinelli; the picture Gilles Deleuze paints of his memory of listening to the boom of a favorite teacher's voice as they walked along a seaside in recitation—that mood of instruction laced with a mood of admiration and affection; and the hard-to-believe scenario of Emily Dickinson belting out "chants, rounds, fugues, and anthems" with a classmate at Mount Holyoke, who recounts the episode in which the two of them flee campus in search of a sequestered and broad "spaceway bounded by the horizon," singing book in hand. We're to picture the two girls—one destined to become America's most original and reclusive poet—"perched upon the topmost rail of a fence" where they "opened the book and [their] mouths, drew the diapason stops of [their] vocal organs, and sang tune after tune,—long metres, short metres, hallelujah metres, et id." Two or three cows quietly feeding in a pasture nearby are their only visible auditors; they "stood in silent amazement of the unusual sight and sound." "We needed no plaudits, for we were a joy to ourselves," Dickinson's mate in truancy, Amelia D. Jones, class of 1849, wrote. "We had found a remedy for depression, repression, suppression and oppression, and no two maidens returned that day from open-air exercises more exhilarated than we." Dickinson is the voluminous pineapple in this still life; Farinelli, the shapely pear; Deleuze the steadying apple.

Writing is a type of arrangement for voice and reader. Writers and readers can never get enough of voice. Compelled to read, needing at all costs to write while dinner burns on the stove and babies cry their eyes out to be fed, it's not a story with its twists of suspense that draws us perpetually back to the page or keeps us in its thrall. We're driven first and foremost by a voice addiction, hopeful that *this time* we'll find the tune to match the mood of what we hear or that, *this time*, we'll be seduced by a tonal range as strange as it is familiar, and as familiar as it is strange. All writing, when it comes down to it, is nothing more and nothing less than the patient re-creation of a voice-filled room.

The writing voice I've always claimed to aspire to—my voice imago—is characterized by a purity, sparseness, and minimalism that I hear in the prose of Natalia Ginsburg, and the holy, ecstatic, quiet, rapturous voice of Sarah Orne Jewett. These qualities are so far afield from my writing voice as I know it and the mood hues I've come to create, in turn, as to be laughable. For, putting to one side my speaking voice, which I experience as husky and naive, I seem to tend toward a writing voice layered and baroque, obdurate and convoluted, blunt and fumbling and blind. Sometimes I "hear the music" in my prose—like most writers, I'm lucky if that comes through sustainably in thirty pages, three sentences, three words, but more likely three syllables. Think about it: is the song itself of a Joni Mitchell (if we dare allow the analogy to hold) the place where a mood most rings, or is the hard-won song merely the pretense for the three notes she absentmindedly strums as an afterthought, unconsciously called to discover if, in performing the song, she's thrown the guitar out of tune? Is all writing written for this after-song?—the tic and quiver that any musician sounds after the song is over, inaudible even to herself, beneath applause?

For me, only when I channel the thing I've written as its vehicle not as its author in the form of a public reading, do I begin to hear its music; then I feel I can rouse and lull an audience in combination. Only by way of that arduous passage—from mind to page and pen and back through (anonymous) body—can I find myself capable of singing in a way that is unavailable to me in daily life.

My mother's voice is always and ever singing. Sometimes it seems to be practicing scales, limbering up. It can move from the deep growl of a "halloooooo" to the gracefully feminine lilt of a melodious only slightly upward turning, "hell-o-a"? Incapable of a whisper, it is bold and multivalent, like an instrument piped by a virtuoso who can reach across several octaves simultaneously or strike more than one note in tandem. Often bell-like, my mother's voice has something of that "chaotic joy" that sound artist Stephen Vitiello achieved by first isolating and then replaying simultaneously fifty-nine different bell sounds otherwise lost inside the soundscape of New York City, from the tink of a cat's bell collar to the gong of Saint Paul's Chapel's "Bell of Hope." My mother's voice gave me an appetite for poetry early on, and the magnificence of its emotional range maybe made me yearn for something near-as-can-be to silence—minimal—but with a hint of a sensibility traceable to Italy and lost to us in immigration.

Think of Leonora in Verdi's *La Forza del Destino*, a hermit who lived in a cave outside a monastery. She's been supplied by the monks with a bell so that she could summon help if trouble arose. Imagine my mother's father, the southern Italian immigrant writing a novel in Italian based on the opera, and a preoccupation with legacy and fate, the novel lost after his death. The opera is known for a famous duet, "Solenne in Quest'ora," for tenor and baritone, and an aria, "Pace Mio Dio," for soprano. My mother tells me she used to blast that music the way the kids of a later generation blasted rock, but her fervency for opera was borne of an apprenticeship with her father on afternoons when they and they alone, in a house that held six children, a mother, and mother-in-law, would weave an opera cocoon near to an open window whose gauze curtains stirred during such sessions in the breeze that drifted in, fragrant, as my mother recalls it, with my grandfather's driveway plantings.

Could a family mood be haunted by the incursion of a musical term into a sentence like a scoring of the days on a page from my Italian immigrant grandfather's journals? The sentences that lay inside a miscellany of scraps of writing that he daily penned in his shoe repair shop read: "The same adagio prevails with us. No change in our behavior." By adagio did he mean torpor, mournfulness, a sloweddown spell that was making it impossible for someone or -ones to act, and thereby change some circumstance or set of relations they'd become trapped by or inured to? "The same adagio prevails with us. No change in our behavior." Slow movement can also be serene, but here he seems to be thinking palsy.

We turn to art to be, in one sense of the term (affectively), moved, and in another sense to be taken out of ourselves so as to move out of the place we are in. "It was good to let peace keep pace with you" was another of his aphorisms, which could be an antidote to an even less likely phrase, "the same presto prevails," "the same pizzicato," "the same allegro prevails." How does a mood come to prevail? By what drive or persistence of vision? Might, or the refusal of might? Will? Color? Should we also speak of an adagio of color? Light has everything to do with duration. Or of smell? Some scents are sharp and short, while others have the capacity to linger and expand. Adagio of loves or intentions, adagio of dreams, adagio of words overheard—the first syllable we ever heard, followed by the many.

My grandfather had brought a collection of 78 LPs with him from Naples when he emigrated. In his more flush years, he listened—on a Victrola that had been a 1922 wedding gift—to Francesco Tamagno, the first of Verdi's Otello's singing in 1898, or the full course of *Cavalleria Rusticana*; he played Titta Ruffo and Amelita Galli-Curci singing "Caro Nome" from *Rigoletto*, or "The Mad Scene" from Donizetti's *Lucia di Lammermoor*; he worshiped at the altar of the voice of Caruso. As my mother would say, her father loved music—she never said, "very much," but "very very very much" in case you might not understand just how fierce the attachment was: as if it were the key to a family vault replete with secret stores of feeling, moods for which there were no names but whose heft and sweep could explain everything, everything lost, left behind, unfulfilled, everything broken when a strange language entered in, everything that was meaningful in a world caught in a maelstrom of stupidity and store-bought splendor. Seared into familial memory was the day on which my grandfather, with the help of his only son, carried the Victrola step by step, draped in a black cloth, to the basement. My grandfather had lost everything in the Depression, and unable to afford repair of the Victrola, which had also failed, he retired the machine along with the record collection to the basement as if to seal its fate alongside the coal bin that from then on was more often a harbinger of bitter cold than warmth.

In the course of ten ensuing years, my mother's brother's hair grew into a thick pompadour. It was like a detail from a fairy tale. Whenever he was caught in worry or deep in thought, he could be found twirling a lock of his lush and wavy hair. Having secured a well-paying job at a bank just out of high school, he worried his hair, hopeful that the answer would come to him of whether there was something he could buy for his father to reduce his misery. His first thought was an overcoat to replace my grandfather's characteristically shabby one. My grandfather rejected the offering, not wanting to play a part that didn't befit the wearer. A few more twirls, and my uncle arrived at the better answer of what they came to call "the recording machine"—a stereo console record player with radio built into the front that would lead to what by now felt like an unearthing of an archive, the retrieval of the records step by step by step; the setting of the console upon a little serving table with double doors that seemed just the right height for it; the installation of the entire

ensemble athwart a window in the dining room; and the invitation
to my mother, a twelve-year-old, to listen if she would, and learn: to
receive the greater gift of the breaking of a silence, and the creation
of an envelope of mood and sound.

After school or on weekends, this became my mother and grandfather's secret share: for each record, he would convey the entire
story of the opera, mostly Italian and French. "O Rachel when the
grace of the Lord turns you over to me," she sings it seventy years later
to the day. She sings "L'African," and "O Paradiso," explaining that
the difficulty of the aria meant the full opera was rarely performed.
You have to realize that my mother remembers each and every detail of these sessions. Absolutely nothing was lost on her, especially
the lessons her father conveyed by way of Caruso, who put his heart
and soul even into an aria whose verse is frivolous; who held back
in order to make a tender point as in "Una Furtiva Lagrima" from
Donizetti's *L'Elisir d'Amore*; or who would deliver a difficult series
of notes without taking a breath to interrupt the phrase; who would
ultimately be revered for entering high note territory where others
feared to tread. You have to picture it—my mother explains—how
the city of Philadelphia was required to build a performance space
large enough to hold the crowds who would come to hear Caruso.
Does the place still stand? I've definitely never heard of it—it was
called The Met (not to be confused with its New York cousin), and
following Caruso's death in 1922, it became a boxing arena. Perhaps
the most telling detail was the instance of crossed paths: when my
Neapolitan and Sicilian grandfathers first met, they discovered only
one thing in common between them: they both had been to the same
performance on the same night of a concert by Caruso at this venue.

When they ran out of 78s to play—though repetition was not
pooh-poohed—my grandfather would send my mother into the city
to purchase particular records for their ongoing afternoon listenings-
in-kind: Paganini's concerto in D Major; the overture to the *Barber
of Seville*. As my mother got older, and approaching her senior year
in high school, she was offered a clerical job in an office next to my
grandfather's shoe repair shop and home. A fellow artisan named

Louie the tailor had long since moved out, and the new tenant
needed additional secretarial help. He was an Armenian physicist
named Dr. Lucier who made helicopter parts and who was, behind closed doors, working to develop a battery-operated watch. My
grandfather was opposed to my mother taking the job—who knows
why—perhaps in that way he had of wanting to keep his children secure in the magic circle he hoped he was creating with his music and
his words. My mother took the job in spite of her father's wishes, but
with one stipulation: that she be allowed to listen to the Metropol-
itan Opera on the radio on Saturday afternoons while she worked.
Her new employer, interestingly, complied.

I am convinced my mother knows the words to every song ever
sung, especially those that were broadcast over radio from the 1930s
through the 1950s. She doesn't just know the first verse of so many
songs of yore—from folk to pop, and then some; of opera scores
galore—she knows the second and third as well, even those songs,
she says, that you wouldn't believe are songs, the lyrics are so stupid.
Whether you ask her or not, she will sing them. She will tell you
the plot of an opera from which an aria has been culled even if you
haven't called for it. She will sing and keep on singing. She has sung
her way through life. She has made of her life a musical no matter
her own occasional demolishment by mood, her own paralysis by
fear, no matter the desperation of a circumstance or the indignities
of age.

If she cries easily, as if on cue, is it because of her share in the
lesson of her father's, the immigrant's lament? Thank goodness for
the breeze in the forging of that envelope they made of sadness and
of passion intertwined.

My mother is recounting to me the time together with her father and referring to the title of a little performed aria. She pauses
before she sings a verse from it to ask, "Do you know the melody?" I
want to laugh. "How the heck would I know the melody, Mom? I'm
not where you were when you were listening, am I? In that circle of
sound. I hadn't yet been conceived when you sat by the console with
your father of a blooming afternoon. Did you sing it to me in utero?

Should I know the melody because you sang it, but I failed to listen long after I was born?"

I never learn the opera that my grandfather taught my mother. I do not take to it in a serious or assiduous way. In fact, a great deal of opera leaves me cold. Which is as about ridiculous a thing to say as the sentence an acquaintance once summarily pronounced that went, "I hate Rome," as if Rome were one thing; as if Rome were something you could have an opinion about; as if Rome cared whether you loved or hated it—it didn't need your affection, your indifference, or your disdain to continue monumentally and in all its singularity to influence and thrive.

When my mother recounted the stories of great operas to me in my childhood, I'd grow bored; I'd twiddle my thumbs and look out the window. There's so much that I've not yet learned most literally. There are familial mood haunts that I've failed to investigate or know. But wait: if I don't talk the opera talk, that's only because I walk the opera walk. See how my skin crackles when it's touched like a needle to the groove of a hundred-year-old recording. Do I know the melody? Of course I don't. It's the envelope of my contentment and my pain. I strive to hit a high note in its midst, listening, true. I want to sing with you. To find a language for a purity of line before my life is through, in step with the pace of an Italian *passeggiata*. To find the means for holding back the voice to make one tender point.

ACKNOWLEDGMENTS

The composition of *Life Breaks In* was a circuitous and unpredictable journey, and I have many people and entities to thank. A 2011 Fellowship in Creative Arts/Nonfiction from the John Simon Guggenheim Memorial Foundation enabled me, in the first place, to travel to Vancouver, British Columbia, where I had the opportunity to deepen my knowledge of sound studies in the place where the World Soundscape Project was birthed. There, I was able to enjoy the influence of composer Barry Truax and his students at Simon Fraser University; to experience inspired "sound walks" with soundwalk artist Andra McCartney; and to learn from visiting cultural theorists like Steven Connor while immersing myself for several months in electroacoustic music, sound atmospheres, experiments, and conferences on sound.

Four years forward of those experiences, a Berlin Prize afforded me time in residence at the American Academy in Berlin where, as the fall 2015 Holtzbrinck Fellow, I was able to bring this manuscript, long in the making, to completion in a setting that spurred and challenged me, by turns. The academy supported my interest in the interdisciplinary creation of "mood rooms" and generously hosted a multimodal performance based on this book, "Of Mood: An Atmospheric Reading." The chance to discuss my ideas with such figures as filmmakers Volker Schlöndorff, Ute Aurand, and Robert Beavers and with psychiatrist Regina Casper was inestimable, as was the creation of an atmosphere congenial to ideas made possible by the academy's staff, including Carol Scherer, Christina Wölpert, Simone Madore, Marie Christine Mitzlaff, John-Thomas Eltringham, Ashley L. Baron, R. Jay Magill, Johana Gallup, Yolande Korb, and Lauren Fritzsche, who assisted me in the preparation of the manuscript. I want especially to acknowledge Kerstin Apel for spending an afternoon in the library talking with me about *Stimmung*, Kleist, and the *biergarten*; Lorraine Daston, distinguished historian of science and director of the Max Planck Institute, whose keen interest and whose thinking on clouds and on the imagination, yielded the perfect image and conceptual frame for the book entire: the cyanometer; and, Carol Scherer and jazz trumpeter Paul Brody who invited me into all manner of Berlin soundscapes, central and fringe, opening me to worlds that could spark new and future writing.

I'm also grateful to those who hosted me in related talks and readings in Germany, in particular, Andrew Gross at the University of Göttingen; Christy Hosefelder who made me aware of artist Smilde Berndnaut's "nimbus series"; Christoff Holzhey and Claudia Peppell at the incomparable Institute for Cultural Inquiry; Laura Scuriatti at Bard College Berlin; and Amir Naaman at Topics Books in Neukölln. All of my fellow fellows at the academy inspired me, and I'm particularly grateful for the kind company, example, reading suggestions and support of Jason Pine, Anthony Marra, Kappy Mintie, Vladimir Kulíc, Christina Schwenkel, Moishe Postone, Christine Achinger, and Philip Kitcher. *Life Breaks In* also enjoyed the conversations, encouragements, and hospitality of Ali Hyman Wolff, Babette Tischleder, Rebecca Loyche, David Moss, Doris Schnelle and Rolf-Dieter Schnelle, Christine Wallich, Dr. Hans-Michael Giesen and Almut Giesen, Cecile Niemitz-Rossant, Melinda Harvey, Anna Weitemeyer, Kaye Mitchell, Daniel Merkel and Liz Rosenthal, Andrew Yang and Christa Donner, Marica Bodrožić and Gregor Hens, Amy Benfer and Tom Drury, Lance Olsen and Andi Olsen, Kenny Fries and Mike McCullough. I am especially grateful to poet and literary translator Alexander Booth for his engagement in an interview we carried out together for NPR Berlin and for conversations still to come.

For the most hassle-free relay of images I have ever experienced, I thank David Rosado of the New York Public Library. For their wondrously swift, thoughtful and efficient responses to my queries on the life and work of Florence Thomas, I thank Markrid Izquierdo, Information Services, Multnomah County Library, and her staff, who provided me with study images and "alice in motion" footage of Thomas's *Alice in Wonderland* bas-relief, as well as hard-to-find newspaper articles on Thomas; Jayson Colomby, who took the photograph of the bas-relief that appears in this book; and, View-Master collectors, documentarians, and experts, Wolfgang Sell and Mary Ann Sell, who supplied the photograph of Florence Thomas at work.

Falling down a Charles D. Hubbard rabbit hole led me to meet numerous amazing people and incomparable guides: Fawn Gillespie, East Haven Memorial Library; Patricia Baldwin, assistant director, Guilford Free Library, the Edith B. Nettleton Historic Reading Room; Guilford Keeping Society members Carl Balestracci, Patricia Lovelace, museum director/curator of the Guilford Keeping Society, and Joel Helander, municipal historian. I owe the most singular gratitude to Deborah Staber, museum director, L. C. Bates Museum, and her assistant, museum educator Serena Sanborn. The continuance of this very special place is the result of a rare and single-minded devotion, and my work on Hubbard, Hinckley, and what I call the "mood rooms" in their care was made possible by Deborah Staber's unusual expertise, openness, and generosity.

Two journals debuted excerpts from *Life Breaks In* in the fall of 2015: thanks to Mary Rockcastle, editor-in-chief of *Water~Stone Review* for featuring "Mood Rooms" as the journal's annual Meridel LeSueur Essay, and R. Jay Magill and Johana Gallup for choosing "Gong Bath" for inclusion in the *Berlin Journal*.

At the University of Rhode Island, my writing and research would not have been possible without the ongoing help of librarian Tawanda Maceia; English Department chair, Ryan Trimm; administrative assistants Michelle Caraccia and Kara Lewis; Dean Winifred Brownell, Provost Donald DeHayes, President David Dooley; the Center for Humanities, and the University of Rhode Island Foundation. For helpful reading suggestions in their respective fields, I thank my English Department colleagues, Jennifer Jones, Stephen Barber, Carolyn Betensky, Valerie Karno, and graduate student extraordinaire, Sarah Kruse; writers Robert Leuci, and Jody Lisberger; philosophers Cheryl Foster and Galen Johnson; and scholar/writers Martha Elena Rojas, Wendy S. Walters, and Patricia Ybarra for their invitation to be part of "The Essay in Public Symposia" at Brown University.

Life Breaks In benefited from the inspired and generous suggestions of so many people, some of which made their way directly into these pages, all of which paved this book's avenues of exploration. From the outset and throughout, or at particularly important junctures, my thinking was aided by the input and support of Ben Tyler; Peter Eudenbach; Adria Evans; David Lazar; Amy Hoffman; Roberta Stone; Peter Covino; Hillel Schwartz; Barrie Jean Borich; Ames Hawkins; Elaine Sexton; Kristen Prevallet; Patrick Madden; Elizabeth Brady; Edvige Giunta; Deb Rosenberg; Brown University School of Medicine faculty member Teresa Schraeder, MD; Brown University Medical student Tamara Feingold; University of Rhode Island undergraduate Hayley Hutchins; Gabrielle Wellman; Ayad Akhtar; and Alison Bechdel.

This isn't the first book I've written that has been enriched by the friendship of Laura Lindgren and Ken Swezey, who lend me rare books, who are always interested, and who have a knack for making the essential connection. If my previous book, *Swallow*, enjoyed the coinspiration of the photographs of Rosamond Purcell, I should have anticipated that Rosamond's eye would also have preceded mine in the uncanny chambers of the L. C. Bates. Rosamond and Dennis Purcell's generous contributions to this book are unsurpassed.

Two anonymous readers gave this book their expert care and attention. Their enthusiastic insight and guidance came at just the right time, as did the receptivity of the impeccable staff at the University of Chicago Press: Editorial Director Christie Henry; Senior Manuscript Editor Yvonne Zipter (who graced these pages with a poet's eye); Promotions Director Levi Stahl; and Editorial Associate Gina Wadas. The fact of Christie Henry's having been open to, and in fact, excited by the experimental contours of this book means everything to me, as do the efforts on my work's behalf of my agent, Malaga Baldi.

If the voice of my mother, Rosemary Petracca Cappello, frames this book, the pages in between are odes to mirthful slumber. *Life Breaks In* took its cue from particular children—some of whom are no longer children but young adults—whose instruction and direct influence are evident in these pages: Sophie Walton, Sam Walton, Hayden Weiser, Alisha Berl, Ava Cappello, Lea Rosenberg, Caeli Carr-Potter, and Kolya Markov-Riss. For soundings too numerous to count, I acknowledge my creative accompanists and steadfast friends, fellow artist/writers, Karen Carr, Russell Potter, and James Morrison. My partner, Jean Walton, is the atmosphere in which I daily wander and delight: our life together the ground of this making.

Without quite saying so, this book is dedicated to the memory of Caren McCourtney (1947–2012), who demonstrated how the best mood work happens neither in captivity and obscurity nor in a mood cocoon, but in a world rife with interest, surprise, and the weird creations of other people; who sought out adventure, for and with me, and who sent me things to "spur me on"; who introduced me to Good Will Hinckley and its off-the-beaten path museum.

If there was ever a book I didn't wish to bring to completion while at the same time pining for a way to bring it to a close, it was this one. Mood isn't the sort of subject one moves on from, or whose "book" is ever finished or closed. If mood materializes in these pages, it just as surreptitiously vanishes, for that's its way. Being in mood's nameless company—haunting and surreal—has been a pleasure and an honor, sort of like having unscheduled teas with the minor gods who orchestrate the self and augur our atmospheres. Parting company, one doesn't know entirely where one has been, nor what, exactly, one has imbibed, only that the place and time was oddly graced. I offer this book so that it might facilitate new recognitions of mood, its transformative ground, and the strange beauty of its ever-present mystery. It is my hope, in concert with others, on and off these pages, to incite new mood realms for our times, indicative of both need and desire, not yet glimpsed, nor easily discerned.

NOTES AND SOURCES

For titles referenced more than once in the course of this book, consult the segment in which the reference first appears for detailed bibliography.

On the Street Where You Live
I wish to thank Anglo-Saxonist, Sarah Higley for helping me to understand "mood's" earliest linguistic roots in English.

Ahuja, Nitin K. "'It Feels Good to Be Measured': Clinical Role-Play, Walker Percy, and the Tingles." *Perspectives in Biology and Medicine* 56 (Summer 2013): 442–51.

Anzieu, Didier. *The Skin Ego: A Psychoanalytic Approach to the Self.* New Haven, CT: Yale University Press, 1989.

Beedie, Christopher J., Peter C. Terry, and Andrew M. Lane. "Distinctions between Emotion and Mood." *Cognition and Emotion* 19, no. 6 (2005): 847–78.

Bennett, Jane. *Vibrant Matter: A Political Ecology of Things.* Durham, NC: Duke University Press, 2010.

Coetzee, J. M., and Arabella Kurtz. "Nevertheless, My Sympathies Are with the Karamazovs: An E-mail Correspondence: May–December 2008." *Salmagundi,* nos. 166–167 (Spring–Summer 2010), 39–72.

Connor, Steven. "The Modern Auditory I." In *Rewriting the Self: Histories from the Renaissance to the Present,* edited by Roy Porter, 203–23. London: Routledge, 1997.

Deleuze, Gilles, and Claire Parnet. "O as in Opera." In *Gilles Deleuze: From A to Z.* Directed by Pierre-André Boutang. Translated by Charles J. Stivale. Los Angeles: Semiotext(e), 2012. DVD.

Felski, Rita, and Susan Fraiman. "Introduction." In "In the Mood," special issue of *New Literary History: A Journal of Theory and Interpretation* 43, no. 3 (Summer 2012): v–xii.

Guignon, Charles. "Moods in Heidegger's Being and Time." In *What Is an Emotion? Classic Readings in Philosophical Psychology*, edited by Chesire Calhoun and Robert C. Solomon, 230–43. New York: Oxford University Press, 1984.

Gumbrecht, Hans Ulrich. *Atmosphere, Mood, Stimmung: On a Hidden Potential of Literature*. Translated by Erik Butler. Stanford, CA: Stanford University Press, 2012.

Highmore, Ben, and Jenny Bourne Taylor. "Introducing Mood Work." In "Mood Work," special issue of *New Formations: A Journal of Culture/Theory/Politics* 82 (2014): 5–12.

Hollander, John. *The Untuning of the Sky: Ideas of Music in English Poetry, 1500–1700*. Princeton, NJ: Princeton University Press, 1961.

Hu, Elise. "Facebook Makes Us Sadder and Less Satisfied, Study Finds." *All Tech Considered*. National Public Radio, August 20, 2013. http://www.npr.org/blogs /alltechconsidered/2013/08/19/213568763/researchers-facebook-makes-us -sadder-and-less-satisfied.

O'Connell, Mark. "Could a One-Hour Video of Someone Whispering and Brushing Her Hair Change Your Life?" *Slate Magazine*, February 12, 2013. Accessed March 12, 2015. http://www.slate.com/articles/life/culturebox/2013/02/asmr _videos_autonomous_sensory_meridian_response_and_whispering_videos _on.html.

Physics arXiv (blog). "Why Physicists Are Saying Consciousness Is a State of Matter, Like a Solid, a Liquid or a Gas." *Medium*, January 15, 2014. https://medium .com/the-physics-arxiv-blog/why-physicists-are-saying-consciousness-is-a -state-of-matter-like-a-solid-a-liquid-or-a-gas-5e7ed624986d.

Ratcliffe, Matthew. "The Phenomenology of Mood and the Meaning of Life." In *The Oxford Handbook of Philosophy of Emotion*, edited by Peter Goldie, 349–71. New York: Oxford University Press, 2012.

Roquet, Paul. "Ambient Literature and the Aesthetics of Calm: Mood Regulation in Contemporary Japanese Fiction." *Journal of Japanese Studies* 35, no. 1 (2009): 87–111.

Ryssdal, Kai. "Would a Marketplace by Any Other Name Smell as Sweet?" *Marketplace*, July 17, 2014. http://www.marketplace.org/topics/business/would -marketplace-any-other-name-smell-sweet.

Spitzer, Leo. *Classical and Christian Ideas of World Harmony: Prolegomena to an Interpretation of the Word "Stimmung."* Edited by Anna Granville Hatcher. Baltimore, MD: Johns Hopkins Press, 1963.

2 **And what if I said that we are always under an influence.** The sense of mood as an ineluctable ontological backdrop or ground of human consciousness is central to the work of Martin Heidegger—mood as "the

primordial phenomenological characteristic of self-experiencing life"
that "precedes everything that has the character of an act" (Ratcliffe,
"Phenomenology," 357). In philosopher Charles Guignon's summary, for
Heidegger, "We can slip over from one mood into another, but we can
never be free of moods altogether" ("Moods," 235). Even an apparent
lack of mood is a mood. Moreover, moods are not to be understood as
inner or psychic phenomena. Quoting Heidegger, "A mood assails us. It
comes neither from 'outside' nor from 'inside,' but arises out of Being-in-
the-world, as a way of such Being" (Guignon, "Moods," 236). By way of
mood, we "tune in" and "turn on" to things, people, the world (Guignon,
"Moods," 237). Felski and Fraiman note that "[mood] is often used, by
Heidegger and others, to convey an overall orientation to the world that
causes it to come into view in a certain way. Mood, in this sense, is not
optional, but a prerequisite for any kind of intellectual engagement. . . .
There is no moodless or mood-free apprehension of phenomena" ("Introduction," vi). See also Ratcliffe, who suggests that moods are deeper
than emotions — by nature of their inconspicuousness. Emotions' pre-
supposing ground, moods are "part of the background structure of in-
tentionality" responsible for our sense of "the meaning of life." Ratcliffe
offers the following example: "To be able to experience fear, one must
already find oneself in the world in such a way that being 'endangered'
or 'under threat' are possibilities." "When we *have* an emotion, we are
already *in* a situation," he writes, "and, as Heidegger appreciates, this
sense of being there depends upon mood" ("Phenomenology," 350, 354).

8   **Gilles Deleuze, proposed in an interview.** In the famous late-life interview with Claire Parnet, Deleuze stipulated that the conversation
was only to be aired posthumously (Deleuze and Parnet, *Gilles*).

12   **I have a hunch that we may be entering a moodless age, but that's
not the same as a depressive, depressing, or depressed one.** A few
years ago, when Stanford comparative literature specialist Hans Ulrich
Gumbrecht proposed a new way of studying literature that would take
mood as its starting point and ground, he also wondered whether we
might be entering a moodless age. In his view, a computer-age with-
drawal from what he terms "presence" can explain his, or our, attention
to and longing for the production of atmosphere and the experience of
moods understood as encounters with presence, or a form of "aesthetic
immediacy that has gone missing" (Gumbrecht, *Atmosphere*, 7, 12).

13   **To be sure, our language for moody states of mind has become
less precise.** In a recent dialogue, the great South African writer J. M.
Coetzee called for a psychology for our times that would "pay more se-

rious attention to mood," and then he offered his own "rudimentary
psychology of moods," divided into four classes "corresponding to the
four elements: earth, water, fire and air" ("Nevertheless," 66). "People
rarely—in fact almost never—act on the basis of reason: people act on
the basis of impulse or desire or urge or drive or passion or mood, and
dress up their motives afterwards to make them seem reasonable," Coetzee writes ("Nevertheless," 66). And then he goes on to explain that we
tend to treat moods as "inconvenient obstacles," even though there is no
such thing as not being in a mood, and that it is impossible to be "mood
neutral" ("Nevertheless," 68). Arabella Kurtz, Coetzee's interlocutor,
makes the point that contemporary therapy "isolate[es] mood and regard[s] it as a problem rather than a fact of existence," which gets at one
keynote of my project.

14   **"There will be no greening of the economy"** (Bennett, *Vibrant Matter*, xii). Relatedly, Highmore and Bourne Taylor cite Jonathan Flat-
ley's foundational point that "any kind of political project must have
the 'making and using' of mood as part and parcel of the project; for,
no matter how clever or correct the critique or achievable the project, collective action is impossible if people are not, so to speak, *in the
mood*" ("Introducing Mood Work," 12). On making stores smell a certain way to inspire consumer confidence, see Ryssdal, "Would a Marketplace"; on Facebook and mood, see Hu, "Facebook Makes Us Sadder"; for thoughtful accounts and analyses of the Internet-generated
ASMR phenomenon—a, to my mind, dystopic manifestation of a newly
minted form of emptiness—see O'Connell and Ahuja, "Could a One-
Hour Video"; on Japanese "ambient literature," see Roquet, "Ambient
Literature"; on the physics of consciousness, see *Physics arXiv* (blog),
"Why Physicists Are Saying."

15   **The idea developed by French psychoanalyst, Didier Anzieu, of what
he calls the "sonorous envelope."** See Anzieu, *The Skin Ego*, and Steven
Connor's guiding incorporation of Anzieu ("The Modern Auditory I").
Sound studies as of the last decade describes an interdisciplinary field
that seeks to displace, or at least courageously nudge aside, the primacy
of the visual in Western culture. Whether this new turn in the humanities is called an "auditory turn," a "sonic intervention" (as the name
of a conference in 2005 in Amsterdam suggested), whether it goes by
the name of "sound studies"—or what I call it: "literary acoustics"—it
has the potential radically to alter the nature of the literary object as
such. Imagine reconceiving Dickinson's poetics, e.g., in terms of the
acoustics of space that was her homestead (including the sounds of her

brother, Austin, and his mistress, Mabel Loomis Todd, orgasming in the living room just below the room where she wrote); or finding a new context for Poe's famous "The Tell-Tale Heart" by surveying it through the ear piece of the newly invented "stethoscope" and the introduction of a method known as "mediate auscultation"; or discovering in "The Fall of the House of Usher" a soundscape of water(falls) in an age when the discovery of the great Niagara as a force and possibility became a major American signifying event. A link between sound and mood appears almost accidentally in Beedie, Terry, and Lane, where they group themes relative to mood for their "duration, intensity, stability, timing and clarity . . . analogous," they write, "perhaps to the way a sound can be described in terms of its duration, volume, pitch and rhythm" ("Distinctions," 853). Turning to German, Hans Ulrich Gumbrecht reminds us of how the word that translates into English as mood or climate— *Stimmung*—more precisely, indicatively, and richly connects with the German *Stimme*, meaning "voice," and *stimmen*, to tune an instrument. Mood is thus, by way of the German, intimately associated with attunement, music, sound, and, by affiliation, touch. If there are two suppositions that recur in sound studies, it is that touch is the nearest sense to listening and that we never just hear with our ears but with our entire bodies. Specific moods, like specific sounds, "present themselves to us as nuances that challenge our powers of discernment and description, as well as the potential of language to capture them." Felski and Fraiman acknowledge that though "individual moods have often been explored in literary and cultural studies . . . sustained theoretical reflection on mood is still in its infancy" ("Introduction," xii). If a turn to mood is relatively new, investigations of *Stimmung* are not, if we consider the centrality of attunement to Heidegger; Spitzer's book-length study of the untranslatability of the German word *Stimmung* and its cognate ideas across centuries and cultures (*Classical and Christian*); or Hollander's study of mood-mode correspondences: attempts to create equivalencies between musical and poetic modes and listeners' moods (*Untuning*).

Mood Modulations

Connor, Steven. *The Matter of Air: Science and Art of the Ethereal*. London: Reaktion Books, 2010.

Barthes, Roland. "Reading Brillat-Savarin." In *The Rustle of Language*, translated by Richard Howard, 250–70. New York: Farrar, Straus, and Giroux, 1986.

Goldsmith, Kenneth. "Seeding the Data Cloud." In *Uncreative Writing: Managing Language in the Digital Age*, 175–87. New York: Columbia University Press, 2011.

Kristeva, Julia. "Psychoanalysis—a Counterdespressant." In *Black Sun: Depression and Melancholia*, translated by Leon S. Roudiez, 1–31. New York: Columbia University Press, 1989.

Stewart, Susan. "Freedom from Mood." In *The Poet's Freedom: A Notebook on Making*, 53–75. Chicago: University of Chicago Press, 2011.

Vitiello, Stephen. "Intimate Listening." YouTube video, 14:00. Posted by PopTech, June 6, 2011. Accessed May 13, 2013. https://www.youtube.com/watch?v=mTzmJRg8Yyo.

30   **Where there is food, mood cannot be far behind.** Mood is implicated not only in our eating of food but in the way we grow food, prepare it, display and arrange it, to say nothing of whether these processes are pursued together or alone. Yet, I am especially moved by the way that Roland Barthes formulates the eating of food as a separate sense, capable of effecting a mood of well-being: "Gustative delight is diffuse, extensive to the entire secret lining of the mucous membranes; it derives from what we should probably consider our sixth sense . . . and which is *cenesthesia*, the total sensation of our internal body . . . the sense of *well-being* which follows good meals."

33   **In the family of origin mood space, did I ever—and if not, why not?—listen to voices by putting my ear to the table, or the wall, or the floor?** These questions are inspired by an illustrated talk by sound artist, Stephen Vitiello ("Intimate Listening").

36   **If moods are the invisible worlds we live in, we might have to trick mood the way that scientists did with air** (Connor, *Matter of Air*).

39   **In the case of certain moods, the minute you start to think about them, they cease to be moods.** See Susan Stewart's brilliant and learned meditation on poetic moods in *The Poet's Freedom: A Notebook on Making*, where she writes: "We speak of being 'in a poetic mood,' yet like any mood, a poetic one is absorbing, and the only certain thing we can know about it is that it is not certain we can 'know' it. We fall into, awaken from, or 'snap out of' moods: if we can write about a mood, we are out of the mood" ("Freedom," 53).

41   **Kenneth Goldsmith calls Facebook status lines "mood blasts"** (Goldsmith, "Seeding the Data," 175).

Of Clouds and Moods

Alcott, Louisa May. *Moods.* Edited by Sarah Elbert. New Brunswick, NJ: Rutgers University Press, 1991.

Broglio, Ron. "From Sky to Skyscape." In *Technologies of the Picturesque: British Art,*

Poetry, and Instruments, 1750–1830, 129–58. Lewisburg, PA: Bucknell University Press, 2008.

The Cloud Appreciation Society. "Cloud Lovers, Unite!" http://cloudappreciation society.org.

Connor, Steven. "Obnubilation." *The Essay*, BBC Radio 3, February 25, 2009. http://stevenconnor.com/obnubilation.html.

Hamblyn, Richard. *The Invention of Clouds: How an Amateur Meteorologist Forged the Language of the Skies*. New York: Farrar, Straus and Giroux, 2001.

Jacobus, Mary. "Cloud Studies: The Visible Invisible." In *Romantic Things: A Tree, a Rock, a Cloud*, 10–35. Chicago: University of Chicago Press, 2012.

Leslie, Charles Robert. *Memoirs of the Life of John Constable, Composed Chiefly of His Letters*. London: Longman, Brown, Green and Longmans, 1845.

Mitchell, Joni. "Both Sides Now (Live 1967)." YouTube video, 5:59. Posted by MusicForYourFunk, June 13, 2009. Accessed July 12, 2013. https://www.youtube .com/watch?v=kTpPHPgzjJo.

Robertson, Lisa. *The Weather*. Vancouver: New Star Books, 2007.

Schuyler, James. *James Schuyler*. Readings in Contemporary Poetry, No. 9. New York: DIA Art Foundation, 1988.

Smailbegović, Adac. "Cloud Writing: Describing Soft Architectures of Change in the Anthropocene." In *Art in the Anthropocene: Encounters among Aesthetics, Politics, Environments and Epistemologies*, edited by Heather Davis and Etienne Turpin, 93–108. London: Open Humanities Press, 2015.

Sterne, Jonathan. "Techniques of Listening." In *The Audible Past: Cultural Origins of Sound Reproduction*, 87–136. Durham, NC: Duke University Press, 2003.

47 **"Begin afresh in the realms of the atmosphere"** (Robertson, *weather*, 10).

47 **Students of romanticism, in particular, . . . have much to say about clouds.** I am drawing upon the work of Ron Broglio, Richard Hamblyn, and Mary Jacobus throughout. Hamblyn quotes Constable's self-appellation in a letter to his friend John Fisher in November 1823: "You can never be nubilous . . . for I am the man of clouds" (*Invention of Clouds*, 309).

47 **Clouds are "protean" and "ever-changing"; they are "ineffable and prodigal forms."** "The primary challenge for Constable became how to master clouds' protean forms and so capture the ever changing weather" (Broglio, "From Sky,"129). "It is an hour to be remembered by historians and daydreamers alike, for by the end of his lecture Luke Howard, by giving language to nature's most ineffable and prodigal forms, had squared an ancient and anxiogenic circle," and, "But the naming of

clouds was a different kind of gesture for the hand of classification to have made. Here was the naming not of a solid, stable thing but of a series of self-canceling evanescences. Here was the naming of a fugitive presence that hastened to its onward dissolution. Here was the naming of *clouds*" (Hamblyn, *Invention of Clouds*, 6, 171). "Clouds are not solid but they are (literally) heavy, owing to their water content," and Jacobus's quoting of Ruskin: "At once material and immaterial, Ruskinian clouds have a way of 'mixing something and nothing' that for him requires explanation" ("Cloud Studies," 26, 21).

48   **"The names for the clouds which I deduced from the Latin are but seven in number."** (Broglio, "From Sky," 146–147, quoting Howard's 1818 *The Climate of London* in two volumes).

48   **I think of poor Laennec, Howard's contemporary and inventor of the stethoscope.** For a full-bodied account of the path that determined Laennec's invention as a culturally and medically significant technique of listening, as well as a discussion of the trials, tribulations, and ultimate failure of his taxonomies, see Sterne ("Techniques of Listening").

48   **Richard Hamblyn . . . recounts how Friedrich resisted** (Hamblyn, *Invention of Clouds*, 307).

49   **A relationship inheres between clouds and human consciousness.** Descartes, Hamblyn reminds us, saw clouds, as "extreme manifestations of the ungraspable" and therefore as "aids to philosophy": "If you can philosophize about the clouds, [Descartes] felt, then surely you can philosophize about anything" (41). Mary Jacobus writes that clouds lend themselves to philosophy, citing the affinity between Merleau-Ponty's thinking on "the visible invisible" and clouds (*Romantic Things*, 12). Steven Connor describes clouds as "having presented themselves to thought as the visible form of thoughts themselves"; clouds "tutor our imaginations," he notes, and focuses his own study on the affiliation of clouds with monstrosity, disease, and their at root etymology, not airiness but "clots," and "clods" ("Obnubilation," 7, 1, and 3).

49   **John Constable again: rather than write that he was trying, let us say, to depict clouds, he described his practice in a letter, "clouds ask me to do something like them."** "My Dear Leslie," Constable wrote on January 20, 1834, "I have been badly ill since you left England, and my mind has been so much depressed that I have scarcely been able to do any one thing, and in that state, I did not like to write you. I am now, however, busy on a large landscape; I find it of use to myself, though little noticed by others. Still the trees and the clouds seem to ask me to try to do something like them" (*Memoirs*, 250).

50   **"As clouds race toward their own release from form"** (Hamblyn, *Invention of Clouds*, 17).

50–51   **"The ocean of air in which we live and move"** (Hamblyn, *Invention of Clouds*, 350).

51   **How an Englishman should congratulate himself "on belonging to a country of mists and clouds and storms"** (Broglio, "From Sky," 157).

52   **"It is this combination of indeterminacy, space, and interiority"** (Jacobus, "Cloud Studies," 11).

53   **"The first day of winter, or if you want to get technical about it, the first wintry day."** All references to Schuyler are from his diary entries (*James Schuyler: Readings*).

54   **Hamblyn's discussion . . . of Thales of Miletus (ca. 624–545 B.C.)** (Hamblyn, *Invention of Clouds*, 31).

57   **"I like the idea of clouds from both sides and some other things from both sides," singer/songwriter, Joni Mitchell says.** From a 1967 live performance recording (Mitchell, "Both").

62   **Constable lent us the verb "skying" . . . "non-captive balloons."** From John Constable's letter to Archdeacon John Fisher dated October 23, 1821: "I have done a good deal of skying—I am determined to conquer all difficulties and that most arduous one among the rest"; and, "All day long the clouds (especially in September) roll over like non-captive balloons so that one can just lie on the grass and watch them with uninterrupted enjoyment" (Broglio, "From Sky," 131, 133). For accounts of problems that critics found with Constable's early skies, see Broglio ("From Sky," 130–34).

62   **It's "a specificity that exceeds classification."** "Constable supersedes Howard's nomenclature by painting a specificity of clouds that exceed classification" (Broglio, "From Sky," 129).

62   **The "scraps and bits of paper" on which he made his observations on clouds and skies never formed into a lecture as he had intended before he died.** "His intention was to deliver, as the highlight of the series, a lecture on the new science of meteorology. This lecture, however, proved more elusive than most of his ambitions. 'My observation on clouds and skies are on scraps and bits of paper,' he wrote to a friend in December 1836, 'and I have never yet put them together so as to form a lecture, which I shall do, and probably deliver at Hampstead next summer.' But by that next summer, Constable was dead, having suffered from depression and ill health for some time. . . . John Constable on 'The Science of Painting the Sky': one of the great lost lectures of the early nineteenth century" (Hamblyn, *Invention of Clouds*, 317).

62 **Fuseli and others have found a persistent mood in Constable's work.** Hamblyn writes of Fuseli's assessment: "This was more than an image of parochial conservatism, for every canvas in Constable Country carries a threat of oncoming rain" (Hamblyn, *Invention of Clouds*, 319).

62 **Among Constable's "favorite weather stories"** (Broglio, "From Sky," 139).

68 **In search of mood hints, I must let clouds act upon me: open and oblique.** In "Cloud Writing," an essay that is companionate with my writing and thinking here but that I only discovered following the composition of these pages, Ada Smailbegović outlines Luke Howard's "desire in devising a taxonomy of clouds to expose and make available to others the transient flow of relations that constitutes the changes in weather phenomena. For this purpose, Howard devised a form of cloud-writing, suggesting the use of concrete, nearly hieroglyphic marks as indicators of specific cloud types" with the effect of "creat[ing] diachronic and synchronic fields of relations" (102). Smailbegović describes what could be understood as a Howardian aesthetic, applicable as well, I would suggest, to writing mood.

Gong Bath

Barthes, Roland. *A Lover's Discourse: Fragments.* Translated by Richard Howard. New York: Hill and Wang, 1978.

Field, Eleanor Selfridge. "Experiments with Melody and Meter, or the Effects of Music: The Edison-Bingham Music Research." *Musical Quarterly* 81 (Summer 1997): 291–310.

Graduate School of Oceanography and Marine Acoustics, University of Rhode Island. "Discovery of Sound in the Sea." http://www.dosits.org.

Histotripsy Group. "Non-invasive Ultrasonic Tissue Fractionation for Treatment of Benign Disease and Cancer–'Histotripsy'" Biomedical Engineering Department, University of Michigan. http://www.histotripsy.umich.edu.

Kahn, Douglas. *Noise, Water, Meat: A History of Sound in the Arts.* Cambridge, MA: MIT Press, 2001.

Kercher, Sophia. "Sound Baths Move from Metaphysical to Mainstream." *New York Times,* August 15, 2015. Accessed August 15, 2015. http://www.nytimes.com/2015/08/16/fashion/sound-baths-move-from-metaphysical-to-mainstream.html?_r=0.

King, Lauren, Quincy Almeida, and Heidi Ahonen, "Short-Term Effects of Vibration Therapy on Motor Impairments in Parkinson's Disease." *NeuroRehabilitation* 25 (2009): 297–306.

Lanza, Joseph. *Elevator Music: A Surreal History of Muzak ®, Easy-Listening, and Other Moodsong ®.* Ann Arbor: University of Michigan Press, 2004.

Nancy, Jean-Luc. *Listening.* Translated by Charlotte Mandell. New York: Fordham University Press, 2002.

Söderlund, G., S. Sikstrom and A. Smart. "Listen to the Noise: Noise Is Beneficial for Cognitive Performance in ADHD." *Journal of Child Psychology and Psychiatry and Allied Disciplines* 48 (2007): 840–47.

Sollier, Pierre. *Listening for Wellness: An Introduction to the Tomatis Method.* Lafayette, CA: Mozart Center Press, 2005.

Storr, Anthony. *Music and the Mind.* New York: Ballantine Books, 1992.

Thompson, Emily. *The Soundscape of Modernity: Architectural Acoustics and the Culture of Listening in America, 1900–1933.* Cambridge: The MIT Press, 2002.

70–71 **Even in its earliest incarnations, the science of acoustics turned to water as its scribe.** "There has been a longstanding association of water and sound, in observational acoustics from antiquity through Chaucer to Helmholtz and beyond, with the sound of a stone hitting water producing a visual counterpart, which was then mapped back onto the invisible movements of sound waves" (Kahn, *Noise*, 246).

72 **This pang that requires Roland Barthes to halt all occupation he calls "reverberation"** (Barthes, *Lover's Discourse*, 201).

72 **Without the aid of microphones or speakers, the sound of gongs materializes and reverberates in the supine body.** See Thompson's magnificently instructive history of early twentieth-century attempts to control sound architecturally in the era that inaugurated a norm of non-reverberant concert hall sound (Thompson, *Soundscape*). Sound produced in the age of the electrical signal—reliant on microphones and loudspeakers—diminished the signature acoustics of a place in return for a cleaner, neater sound. Though sound production would always be dependent on materials in the room, the aim was for the material to remind a listener of the sound and not of itself as an interfering bulk. I'm suggesting that the gong bath returns sound's materiality to the reverberant surfaces in the room and makes one feel that one's body is the sound's medium.

72–73 **A student of mine once made me aware of prescribable sounds, or "audioceuticals."** Thanks to Hayley Hutchins for introducing me to concepts such as "stochastic resonance," "vibroacoustic therapy," and "audioceuticals" in a stunning annotated bibliography she compiled for my Honors Seminar in Literary Acoustics at the University of Rhode Island, fall 2012. See also Sollier: "The time when modern medicine will prescribe a diet of specific sounds to support our wellness is still in the distant future" (Sollier, *Listening*, 67). For a fascinating history of capitalist dystopic forms of "audioanalgesics" from Muzak—"made and programmed for business environments to reduce stress, combat fatigue, and enhance sales"—to mood music—see Lanza. Lanza points to various forms of neutralizing music meant "to kill the pain of urban din," including the 1921 marketing ploy devised by Thomas Edison to sell his records with what he called the "Mood Change Chart" (Lanza, *Elevator*, 4, 11). Resulting from tests performed by Edison and a psychologist, a thirty-two-page pamphlet emerged that grouped records according to the near-medicinal effects the music could have on one's mood, from "Records 'To Bring You Peace of Mind,' 'To Make You Joyous,' 'For More Energy,' 'For Tender Memory,' etc." Widely circulating ads tried to validate results by featuring William J. Burns as the charts' first subject. A famous head of a detective agency, he was understood to be a person least susceptible to emotion. See Field ("Experiments") for an in-depth account of Edison's program.

73   **Oceanographers tell us that sound moves faster in water than it does in air.** All references to the science of underwater acoustics herein are drawn from the treasure trove that is the University of Rhode Island's Graduate School of Oceanography's instructional website, DOSITS (Discovery of Sound in the Sea).

73   **A philosopher steps in and says the body itself is a skin stretched over resonant matter beneath.** "Timbre can be represented as the resonance of a stretched skin (possibly sprinkled with alcohol, the way certain shamans do), and as the expansions of this resonance in the hollowed column of a drum. Isn't the space of the listening body, in turn, just such a hollow column over which a skin is stretched, but also from which the opening of a mouth can resume and revive resonance?" (Nancy, *Listening*, 42).

75   **Unlike great music "that move[s] us because it is expressive of sadness," not by "making us sad."** From Peter Kivy, *The Corded Shell: On Musical Expression* (quoted in Storr, *Music*, 30).

Sonophoto: Boy Screaming

Anderson, Ben. "Affective Atmospheres." *Emotion, Space and Society* 2, no. 2 (December 2009): 77–81.

Böhme, Gernot. "Atmosphere as the Fundamental Concept of a New Aesthetics." *Thesis Eleven* 36 (1993): 113–26.

Carson, Anne. "The Gender of Sound." In *Glass, Irony and God*, 119–39. New York: New Directions, 1995.

Chion, Michel. *The Voice in Cinema*. Translated by Claudia Gorbman. New York: Columbia University Press, 1999.

Diaconu, Mădălina. "Patina—Atmosphere—Aroma: Toward an Aesthetics of Fine Differences." *Analecta Husserliana* 92 (2006): 131–48.

Heschong, Lisa. *Thermal Delight in Architecture*. Cambridge, MA: MIT Press, 1999.

MedLine. "Changes in the Newborn at Birth." Updated April 12, 2013. http://www .nlm.nih.gov/medlineplus/ency/article/002395.htm.

Putterman, Seth J. "Sonoluminescence: Sound into Light." *Scientific American* 272, no. 2 (February 1995): 32–37.

Sloterdijk, Peter. "The Siren Stage: On the First Sonospheric Alliance." In *Bubbles*, vol. 1 of *Spheres: Microspherology*, 477–521. Translated by Wieland Hoban. Los Angeles: Semiotext(e), 2011.

81   **"We tend to call the woman's cry a scream, and the man's cry a shout"** (Chion, *Voice*, 78–79). See also Anne Carson's bracing essay on gender and voice that begins: "It is in large part according to the sounds people make that we judge them sane or insane, male or female, good, evil, trustworthy, depressive, marriageable, moribund, likely or unlikely to make war on us, little better than animals, inspired by God. These judgments happen fast and can be brutal. Aristotle tells us that the high-pitched voice of the female is one evidence of her evil disposition, for creatures who are brave or just (like lions, bulls, roosters and the human male) have large deep voices" ("Gender," 119). She describes the ancients' tendency to collapse quality of voice and use of voice "under a general rubric of gender. High vocal pitch goes together with talkativeness to characterize a person who is deviant from or deficient in the masculine ideal of self-control" (Carson, "Gender," 119). Women and eunuchs were understood to "have the wrong kind of flesh and the wrong alignment of pores for the production of low vocal pitches" (Carson, "Gender," 120). "In general the women of classical literature are a species given to disorderly and uncontrolled outflow of sound—to shrieking, wailing, sobbing, shrill lament, loud laughter, screams of pain or of pleasure and eruptions of raw emotion in general. As Euripides puts it, 'For it is woman's inborn pleasure always to have her current emotions coming up to her mouth and out through her tongue.'" Even today, Carson notes, a man's voice is not allowed to "have too much smile in it" ("Gender," 126, 120).

84   **We are all reliant on self-propelling mood spheres to carry us through our days.** For an entire philosophy of bubbles, foam, and what

he calls "microspherology," see Sloterdijk, "Siren." I feel as though I am inadvertently translating him here, and being inspired by what could serve as a definition of mood when he writes, "My existence includes the presence of a pre-objective something floating around me; its purpose is to let me be and support me. Hence I am not as current systemists and bio-ideologues claim I think, a living being in its environment; I am a floating being with whom geniuses form spaces" (Sloterdijk, "Siren," 478). See also Mădălina Diaconu's meditations on atmosphere as definitionally analogous with mood ("Patina"), and, in sound scholar Marc Crunelle's formulations, which she quotes, also bubble-like: "An atmosphere is generally emanated by a person or a place and—according to Hermann Schmitz and Gernot Böhme—is never an objectively measurable property, nor the mere subjective projection of the person who feels it. A person's air is some sort of personal aura or 'emanation of the essence' (*Wesensausstrahlung*), and produces an immediate pleasant or unpleasant effect on the perceiver, suggesting trust or mistrust. Marc Crunelle describes it as a sort of bubble (*bulle*) or as an invisible, intimate space, specific to every person" (136). See also Anderson's lucid and provocative attempts to conceptualize atmosphere, and by affiliation, mood, without sacrificing its ambiguity or indeterminacy ("Affective Atmospheres"). Anderson refers us to Böhme's emphasis on the material roots of the word atmosphere—"*atmos* to indicate a tendency for qualities of feeling to fill spaces like a gas, and *sphere* to indicate a particular form of spatial organization based on the circle . . . atmospheres have, then, a characteristic spatial form—diffusion within a sphere" (80).

84   **Contained and cut off by "invisible but impenetrable barriers."** See Ratcliffe's analysis of such in the context of his reflections on the phenomenology of mood: "What is lost [in depression] is not just experience of *actual* connections. Experience no longer incorporates the sense that such connections are possible. This is frequently communicated in terms of an invisible but impenetrable barrier or container that irrevocably separates the sufferer from things and people" (Ratcliffe, "Phenomenology," 359). Ratcliffe draws upon Sally Brampton's *Shoot the Damn Dog: A Memoir of Depression* (2008); J. Hull's *Touching the Rock: An Experience of Blindness* (1990); William Styron's *Darkness Visible* (2001); and E. Wurtzel's *Prozac Nation: Young and Depressed in America* (1996).

85   **Their aim is to achieve "a constant temperature everywhere and at all times."** Heschong offers the example of Los Angeles and Houston, cities in which people move from air-conditioned cars, to air-

conditioned offices, air-conditioned restaurants and movie theaters and home, punctuated by the "brief inconvenience of a blast of hot air between the car and the office," and the attendant fantasy of enclosing such cities in climactic bubbles so that "outdoors" would be a thing of the past (Heschong, *Thermal*, 20).

85 **And proposes a "thermal sense" as its own faculty** (Heschong, *Thermal*, 18, 29).

Is It Possible to Die of a Feeling?

Damasio, Antonio. *The Feeling of What Happens: Body and Emotion in the Making of Consciousness*. New York: Harcourt, 1999.

Davidson, Richard J. "On Emotion, Mood, and Related Affective Constructs." In *The Nature of Emotion: Fundamental Questions*, edited by Richard J. Davidson and Paul Ekman, 51–55. New York: Oxford University Press, 1994.

Kristeva, Julia. "The Malady of Grief: Duras." In *Black Sun: Depression and Melancholia*, translated by Leon S. Roudiez, 219-259. New York Columbia University Press, 1989.

Mellor, David. "The Delirious Museum." In *Museology: Photographs by Richard Ross*, 15–20. Santa Barbara: Aperture, 1989.

Shusterman, Richard. "Thought in the Strenuous Mood: Pragmatism as a Philosophy of Feeling." *New Literary History: A Journal of Theory and Interpretation* 43, no. 3 (2012): 433–54.

95 **From "physiological and neurological through to behavioral and social."** "A significant feature of emotion-mood distinctions in the literature is that none of them, despite their intuitive appeal and complexity, are supported by published data. Even traditionally data-rich sub-disciplines, such as neurology and psychophysiology, appear to make relatively arbitrary distinctions" (Beedie, Terry, and Lane, "Distinctions," 849). Philosopher Richard Shusterman concurs: "The distinctions between the affective terms mood, feeling, and emotion are vague, confused and contested not only in everyday discourse but even among academic specialists in the field" (Shusterman, *Thought*, 438). In earlier work, Davidson had acknowledged that even when such distinctions seem convincing, most "can be found wanting when carefully scrutinized" (Davidson, *On Emotion*, 51).

96 **One can see why Damasio relegated this review of the state of the art of our understanding to a footnote.** Damasio offers "happiness, sadness, fear, anger, surprise or disgust" as primary or universal emotions; secondary or social emotions such as "embarrassment, jealousy,

guilt or pride"; and what he calls background emotions, "such as well-being or malaise, calm or tension" (Damasio, *Feeling*, 341n10, 51).

96   **"Language does not always represent psychological reality," the authors warn.** "Because we are able to *say* that emotion and mood are different does not mean that they are, and any difference may be purely semantic. Therefore, emotion and moods may be different words for the same construct or different words for different constructs" (Beedie, Terry, and Lane, "Distinctions," 847). While it's true that our lexicon for emotional and mental states might be, culturally speaking, impoverished—think about the meaningless ubiquity of words like "depression" and "stress"—rather than conclude that we must therefore turn away from language or consider it an untrustworthy ground of investigation, I suggest that we require ourselves to be more studiously and responsibly playful in its name, allowing it to greet us with the renewed vigor of nuance, capable of availing us of realities not yet glimpsed. I find Kristeva describing what I have in mind here: "Beginning with Heidegger and Blanchot respectively evoking Hölderlin and Mallarmé, and including the Surrealists, commentators have noticed that poets—doubtless diminished in the modern world by the ascendancy of politics—turn back to language, which is their own mansion, and they unfold its resources rather than tackle innocently the representation of an external object" (Kristeva, "Malady," in *Black Sun*, 224).

98   **"A moody person is one whose reactions"** (Beedie, Terry, and Lane, "Distinctions," 850).

The Flower Inclines toward Blue

Calvet, Louis-Jean. *Roland Barthes: A Biography*. Translated by Sarah Wykes. Bloomington: Indiana University Press, 1995.

Cavell, Stanley. "An Emerson Mood." In *Emerson's Transcendental Etudes*, edited by David Justin Hodge, 20–32. Stanford, CA: Stanford University Press, 2004.

Ehrenreich, Barbara. *Living with a Wild God*. New York: Twelve, 2014.

Gass, Willliam H. "Emerson and the Essay." In *Habitations of the Word: Essays*, 9–49. New York: Simon and Schuster, 1985.

Harrison, Thomas J. *Essayism: Conrad, Musil, and Pirandello*. Baltimore: Johns Hopkins University Press, 1992.

Huffington Post. "Hedonometer Happiness Index to Chart Moods around the World." April 15, 2013. Accessed May 2, 2013. http://www.huffingtonpost.co .uk/2013/04/30/the-hedonometer-happiness-index_n_3185142.html.

Kauffmann, R. Lane. "The Skewed Path: Essaying as Unmethodical Method." In

Essays on the Essay: Redefining the Genre, edited by Alexander J. Butrym, 221–40. Athens: University of Georgia Press, 1989.

Levine, Alan M. "Skeptical Triangle? A Comparison of the Political Thought of Emerson, Nietzsche, and Montaigne." In *A Political Companion to Ralph Waldo Emerson*, edited by Alan M. Levine and Daniel S. Malachuk, 223–64. Lexington: University Press of Kentucky, 2011.

Martin, Emily. *Bipolar Expeditions: Mania and Depression in American Culture.* Princeton, NJ: Princeton University Press, 2007.

Morris, William N. *Mood: The Frame of Mind.* New York: Springer, 1989.

Regosin, Richard L. "'*Mettre la theorique avant la practique*': Montaigne and the Practice of Theory." In *Montaigne after Theory, Theory after Montaigne*, edited by Zahi Zalloua, 264–80. Seattle: University of Washington Press, 2009.

Thayer, Robert E. *The Biopsychology of Mood and Arousal.* Oxford: Oxford University Press, 1989.

The Man in Blue. "Mood Map—How Happy Is the World?" Last modified May 19, 2010. Accessed April 20, 2013. http://www.themaninblue.com/writing/perspective/2010/05/19/.

Tompkins, Peter, and Christopher Bird. *The Secret Life of Plants.* New York: Harper Collins, 1973.

101   **Maude of sitcom fame put the phrase "manic depression" into public discourse** (Martin, *Bipolar*, 21).

107–8   **"In the US economic system, there is a premium on measuring . . . arousing what John Maynard Keynes' famously referred to as the market's reliance on people's 'animal spirits'"** (Martin, *Bipolar*, 175, 197).

108   **"In Kraepelin's case, an individual's entire life span could be described on one page"** (Martin, *Bipolar*, 186).

109   **There are lists of pioneers** (Thayer, *Biopsychology*, 16).

111   **Psychologists acknowledge that "mood" enjoys a "history of casual use"** (Morris, *Mood*, 1–2).

112   **"First, precise pharmacological control of all mental symptoms"** (Martin, *Bipolar*, 278).

112   **That an individual may be in a "positive and negative mood at the same time"** (Thayer, *Biopsychology*, 17).

112   **Moods as barometers, monitors, or cueing mechanisms in a "self-regulating system"; or when moods are understood as residues of receding emotion, their more mercurial aftereffect** (Morris, *Mood*, 9, 25).

112   **And what of rhythm?** See Susan Stewart on rhythms' ineluctable and implicit influence—atmospheric, biological, and linguistic—on poetic moods and receptivity ("Freedom").

112–13   **"What the terms mean, how they should be applied, and even whether doctor or patient will get to apply them are all matters of contention"** (Martin, *Bipolar*, 99).

113   **"In the process of coming into being," Martin argues, "the public health crisis in moods has changed the way people experience their moods"** (Martin, *Bipolar*, 193).

114   **Those white-gray figurines so popular in the '70s.** The plastic figurines, known as Silisculpts, were made by R and W Berries Companies, Inc., also known as Russ and Wallace Berries Companies.

116–17   **"A pervasive and sustained emotion that colors the perception of the world"** (Martin, *Bipolar*, 43).

117   **"It is said to be bad for psychological or psychiatric problems, signifying danger"** (Martin, *Bipolar*, 151).

117   **Asked how they deal with bad moods . . . "survey data indicate moods . . . expressive behavior"** (Morris, *Mood*, 104).

117   **Thayer, in *The Biopsychology of Mood and Arousal*, prefers the tried-and-true method of a "short brisk walk"** (Thayer, *Biopsychology*, 24).

119   **Only by essaying mood.** If there is a science to mood, it's philosophy; if there is a form, it's the essay where propositions assert and resist logic, allow for pregnant pauses, and the counterdependencies of aphorism. I am not alone in suggesting that mood wants the essay and that the essay inclines toward mood. Gumbrecht, in proposing that the essay makes way for mood without aiming to capture it, supplies the example of Georg Lukács's *Soul and Form*, to which we might return for a model of essayism in writing mood: "Lukács demanded that essays deviate from the 'scientific' goal of finding the truth. 'It is right for the essayist to seek the truth,' Lukács wrote, 'but he should do so in the manner of Saul. Saul set out to find his father's donkeys and discovered a kingdom; thus will the essayist—one who is truly capable of seeking the truth—find, at the end of his way, what he has not sought: life itself.'" (Gumbrecht, *Atmosphere*, 17). See Thomas Harrison's helpful account of the essay as an "open, self-seeking form—more digressive than systematic, more interrogative than declarative, more descriptive than explanatory. . . . It is anti-generic" (3). See also R. Lane Kauffmann's reflections on the essay's kinship or clash with philosophy and criticism, with system and fragmentation from Lukács through Walter Benjamin and Theodor Adorno. Kauffmann

quotes the famously reiterated statement of Roland Barthes uttered toward the end of his career that he had produced "only essays, an ambiguous genre in which analysis vies with writing" ("Skewed Path," 237). A legacy from Montaigne through Emerson and Barthes yokes the essay historically to mood: the *Essais* are understood to have emerged from a particularly Montaignian mournful mood (then, subsequently, to have been turned to by Nietzsche to improve his own mood; Regosin, "*Mettre la theorique*," 274; Levine, "Skeptical Triangle?" 223); William Gass and Stanley Cavell study Emerson as a "philosopher of moods" (Cavell, "Emerson Mood," 26); and the basis of all of Roland Barthes's writing, according to his biographer, Louis-Jean Calvet, was that he had to transform one of his bad moods into a theory. Barthes's "ontology of mood" as a "favorite theme," *surpassing* the notion of good or bad moods, implicating one's "entire inner being, one's viscera and brain," Calvet explains, coincided with his trying to be faithful to his moods even if writing was understood as a place where one could go to "remonstrate" (*argumenter*) with them (*Roland Barthes*, 84). Writing essays doesn't cure or solve or soothe these writers' moods—or if it does, that's an effect as occasional as the weather. More to the point: essaying enabled them to think with mood, not just about it, and mood, from the start, demanded their attention.

They're Playing Our Song

The solo piano concert referenced here featured Meng-Chieh Liu performing Brahms's Waltzes, Opus 39, at the Curtis Institute, Philadelphia, October 10, 2014. Meng-Chieh Liu's career was launched when, as a twenty-one-year-old student in 1993, he successfully substituted for André Watts at Philadelphia's Academy of Music with three hours' notice. Not long thereafter, his career was interrupted by a rare and debilitating illness of the connective tissues. Following a period of hospitalization and long rehabilitation, he gave the lie to a dire prognosis, returning to the stage and a life in music.

Mood Questionnaire

Weschler, Lawrence. "Museum of Jurassic Technology." *All Things Considered*, National Public Radio, December 6, 1996. http://soundportraits.org/on-air /museum_of_jurassic_technology/.

128 **In a sound portrait of this magical place, Lawrence Weschler.**
Weschler quotes David Wilson: "There are some people who, for reasons we don't entirely understand, come to the museum and in our in-

troductory slide show begin to laugh, and laugh uproariously through the entire museum, laugh at every exhibit. We don't object to this, but we don't exactly understand why. On the other end of the spectrum we have people for whom the museum is a much more solemn and serious place. We remember fondly a man called John Thomas, who when he first came to the museum spent at least three hours in the back, and when he came out he leaned his head against the wall and cried inconsolably for at least three minutes. My wife went over to him, and he said that he realized that it was a museum but to him it was like a church." There is a question of whether affective responses line up with the express interest of the museum in "helping people to achieve states of wonder" (Weschler, "Museum").

128   **For James, there are serious, carefree, easygoing moods.** "Constituting [James's] fundamental attitude to philosophy and to life," Shusterman explains, "the strenuous mood is repeatedly invoked by James, even before he officially identified pragmatism as a philosophy and declared his allegiance to it" (Shusterman, *Thought*, 437). Whether William James substitutes "earnest" or "serious" for "strenuous" in his writing, Shusterman underlines the word's connotation of "vigorous active energy and robust effort" (*Thought*, 436). For a full litany of moods that William James notes without defining, see Shusterman (*Thought*, 439).

129   **If mood defies representation, we should be protective of it.** David Mellor's meditations on Richard Ross's photographs of natural history displays address this idea of mood as a mystery under threat. Indeed, Ross's interpretation of the adult's experience of the uncanny in such museums as based in childhood memories that have been sanitized or displaced contribute to my thinking later in this book on Florence Thomas's View-Master scenarios and the habitat dioramas of Charles D. Hubbard. "As Ross says: '. . . one has childhood memories of what display is and upon seeing these same displays as an adult, they are replaced by the modernization and in fact sterilization and sanitization of the (museum) environment.'" Mellor goes on: "Aura, magic—these are Ross's lost horizons and his project begins to take on the look of a protective, reparative act conserving threatened mysteries: To be sure, Boris Karloff has vanished; hence, Ross's tactic to reinvolve him and to recapture the *Stimmung* and aura in his *guignol* presence. All this he does in order to combat the inauthenticity inherent in the spectacle of a curatorially sanitized display of a child's possessions, 'the objects of Tutankhamen shown . . . in perfect plexiglass boxes, with perfect lights'" (Mellor, "Delirious Museum," 18).

129 **For the pragmatists who founded an American philosophical tradition of that name.** I am paraphrasing Shusterman's deeply informative and fluent essay on the subject ("Thought").

In a Studious Mood

Agamben, Giorgio. "The Idea of Study." In *Idea of Prose*, 63–65. Translated by Michael Sullivan and Sam Whitsitt. Albany: State University of New York Press, 1985.

Ehrlman, Veit. *Reason and Resonance: A History of Modern Aurality*. New York: Zone Books, 2010.

Foucault, Michel. "Prison Talk." In *Power/Knowledge: Selected Interviews and Other Writings, 1972–1977*, edited by Colin Gordon, 55–62. New York: Pantheon, 1980.

Larsell, Olof. "Anatomy of the Ear." In *Diseases of the Nose, Throat, and Ear*, edited by Chevalier Jackson and Chevalier L. Jackson, 291–311. Philadelphia: W. B. Saunders Company, 1959.

Ronell, Avital. *The Telephone Book: Technology—Schizophrenia—Electric Speech.* Lincoln: University of Nebraska Press, 1989.

Walser, Robert. "Six Little Stories." In *A Schoolboy's Diary*, translated by Damion Searls, 45–51. New York: New York Review Books, 2013.

131 **It doesn't begin, "Now is the winter of our discontent made glorious summer by this sun of York."** The opening sentences refer respectively to Shakespeare's *Richard III*; Edgar Allan Poe's "The Fall of the House of Usher"; Herman Melville's *Pierre*; Robert Pinsky's 1994 translation of Dante's *Inferno*; and Virginia Woolf's *Between the Acts*.

132 **It begins, "The ear includes three anatomical dimensions"** (Larsell, "Anatomy," 291).

133 **The doctor-anatomist facing the challenge of the ear's unknowability by creating a language, bony and protuberant.** I'm stirred, hopeful to work in sync with Veit Ehrlman's argument that "despite the recent flurry of studies on 'auditory culture,' the physical ear has maintained a strangely elusive, incorporeal present. In their eagerness to save the sonic from modernity's alleged condescension, students of auditory culture have focused almost exclusively on hearing as a metaphorical construct" (Ehrlman, *Reason*, 17). Ehrlman hopes to offer new paradigms for understanding the physical ear through the social and vice versa.

134 **Certain sentences must be read aloud to glean their full effect** (Larsell, "Anatomy," 291, 294, 300, 308).

135 **"Especially susceptible to frost bite"** (Larsell, "Anatomy," 292).

136   **"Those who study," Agamben writes, "are in the situation of people who have received a shock"** (Agamben, "Idea," 64).

138   **Who "sits in a low-ceilinged room 'in all things like a tomb'"** —
Agamben quoting Melville in "Bartleby" (Agamben, "Idea," 65).

138   **"The antero-inferior quadrant is strongly illuminated"** (Larsell, "Anatomy," 294).

Mood Rooms

Deleuze, Gilles, and Claire Parnet. "A as in Animal." In *Gilles Deleuze: From A to Z.*
Directed by Pierre-André Boutang. Translated by Charles J. Stivale. Los Angeles: Semiotext(e), 2012. DVD.

153   **The Ear of Dionysius (*Orecchio di Dionisio*).** Presumably coined by
Caravaggio, the Ear of Dionysius doesn't have the Greek god of pleasure
in mind but a tyrant king who was said to use the acoustically resonant
cave as a prison panaudicon the better to overhear prisoners at any time
of night or day.

154   **French philosopher Gilles Deleuze summarizes** (Deleuze and Parnet,
Gilles).

Miniature Verandas and Voluminous Velvet Forms

I wish to acknowledge affinities discovered late in the composition of these pages
with Lia Purpura's beautiful essay, "Sugar Eggs: A Reverie," whose associative
chain moves in an uncannily companionate way, but in her case via the example
of View-Masters, miniature sugar eggs, natural history dioramas, and Manhattan
window displays.

"April 3, 1957, Jacqueline Kennedy on 'Home' hosted by Arlene Francis." YouTube
video, 9:10. Posted by HelmerReenberg, November 11, 2012. Accessed June 1,
2014. https://www.youtube.com/watch?v=XebG54eybsc.

Bollas, Christopher. *The Shadow of the Object: Psychoanalysis and the Unthought
Known.* New York: Columbia University Press, 1987.

Boxx, Eleanor. "Sculptor, Set Designer 'Changing Scene.'" *Oregon Journal* (Portland, OR), March 1, 1971.

Department of Media, Culture and Communication, New York University. "Dead
Media Archive." Last modified fall 2010. http://cultureandcommunication.org
/deadmedia/index.php/Viewmaster.

de Lauretis, Teresa. "Queer Texts, Bad Habits, and the Issue of a Future." *GLQ: A
Journal of Lesbian and Gay Studies* 17 (2011): 243–63.

"Eames Lounge Chair Debut in 1956 on NBC." YouTube video, 11:21. Posted by

Omidimo, April 18, 2011. Accessed June 1, 2014. https://www.youtube.com /watch?v=z_X6RsN-HFw.

Houzel, Didier. "The Concept of the Psychic Envelope." In *Psychic Envelopes*, edited by Didier Anzieu, 27–58. London: Karnac Books, 1990.

Munn, B. K. "Unsung Geniuses: Florence Thomas of Viewmaster." *Mystery Hoard: Comics, Culture, and Class* (blog). Last modified April 24, 2008. Accessed February 15, 2014. http://frequential.blogspot.com/2008/04/unsung-geniuses -florence-thomas-of.html.

Purpura, Lia. "Sugar Eggs: A Reverie." In *On Looking: Essays*, 35–52. Louisville, KY: Sarabande Books, 2006.

Schwab, Gabriele. "Words and Mood: the Transference of Literary Knowledge." *SubStance* 26 (1997): 107–29.

Sell, Mary Ann, Wolfgang Sell, and Charley Van Pelt. *View-Master Memories*. Maineville, OH: Mary Ann Sell and Wolfgang Sell, 2000.

Timberg, Bernard. "Arlene Francis." Museum of Broadcast Communications. Accessed June 1, 2014. http://www.museum.tv/eotv/francisarle.htm.

"View-Master Early 1946 Fairytales." YouTube video, 14:51. Posted by clementj1, June 6, 2013. Accessed June 10, 2014. https://www.youtube.com/watch?v =YExHNobzkpo.

"View-Master NSA Portland 89." YouTube video, 7:40. Posted by clementj, December 29, 2013. Accessed June 10, 2014. https://www.youtube.com/watch?v =OroJ32IqkdE.

"View-Master Sam Sawyer." YouTube video, 14:44. Posted by clementj1, June 6, 2013. Accessed June 10, 2014. https://www.youtube.com/watch?v =TWboRNN7Mwc.

165–66 **Starting in 1946 and continuing through the 1960s.** Actually, according to a more definitive source, *View-Master Memories* by Mary Ann Sell, Wolfgang Sell, and Charley Van Pelt, Thomas began work for View-Master part time in 1944, at which time she mostly worked on repairing figures made by artists based in Hollywood. In 1946, she started work full time, designing and developing her own "tabletop" stories, beginning with a wildly popular View-Master of *Rudolf the Red-Nosed Reindeer*. According to the Sells, "The unique talent of Florence Thomas produced hundreds of miniature sculptures hand made out of red clay. Her sculptures have become the essence of View-Master tabletop sets." All told, from 1946 until her retirement in 1971, Thomas "crafted more than 80 View-Master reels comprising more than 560 different sets" (Sell, Sell, and Van Pelt, *View-Master*, 93, 94). An average reel sold over ten thousand copies per month (Sell, Sell, and Van Pelt, *View-Master*, 115). Some of the challenges met by Thomas's painstaking work included the necessity to produce an "exact likeness" of certain characters from scene to scene and set to set while also satisfying, in some cases, the original comic book creators whose characters such figures were copying, and to craft the figures in such a way to allow for shrinkage in drying as they were fit into each scene (Sell, Sell, and Van Pelt, *View-Master*, 96). *View-Master Memories* is the most all-inclusive documentary account of View-Master history and ephemera available. For a lively and helpful media-based history of the device, see New York University's online "Dead Media Archive" (Department of Media, Culture and Communication).

168   **Why Clement translated the View-Master into a 2-D YouTube slide show.** According to various Internet sources, Mattel is currently partnering with Google to market a kid-friendly, immersive, digital product reminiscent of View-Master but with gaming and virtual reality elements built in. It's unclear as of yet whether the product will sail or flop. As well, not too long ago, DreamWorks Pictures announced its intention to buy the movie rights to View-Master. Though the film was slated to appear in 2012, it never got off the ground, perhaps failing at that moment when the writers considered having the characters teleport to another world. View-Master is being turned to as a commodity worth exploiting and renewing; the question is whether its specific, material features and consequent mood-creating capacity are adaptable to the ways of viewing and of being that mark our current age.

169   **You can imagine the fun they have with metaphor as the ego becomes ever more abstract** (Houzel, "Concept," 45).

170   **That "during her sessions, [the analyst] left by a concealed door"** (Houzel, "Concept," 46).

170   **What Bollas compellingly calls the "unthought known"** (Bollas, *Shadow*, 4).

170   **Such "untranslated residues," psychoanalytically speaking.** See de Lauretis's helpful précis of, in particular, French psychoanalyst, Jean Laplanche's account of the untranslatable residues that inform adult sexuality. I am suggesting that, in the following quotation, what is said of sex and sexuality can also be said for mood:

> Succinctly stated, Laplanche maintains that sexuality is not innate or present in the body at birth but comes from the adult other(s). . . . It is implanted in the newborn infant—a being without language (as the etymology of the word *infant* implies) and initially without an ego—by the necessary actions of maternal care, feed-

ing, cleaning, holding and so on, through the enigmatic messages they transmit, enigmatic not only because the baby is not able to translate them but also because they are imbued with (un)conscious sexual fantasies of the adult(s), parent(s), or caretaker(s). Untranslatable, these enigmatic signifiers are subjected to primal repression and constitute the first nucleus of the child's unconscious, the primal unconscious. Partial translations occur as the child grows and the ego is formed and develops, but they, too, leave untranslated residues that live on in the individual's mental apparatus as the unremembered memory of bodily excitations and pleasures. Such unconscious memory traces act, in Laplanche's words, "like a splinter in the skin," or we might say, like a virus installed in a computer. They remain live, though undetected, and are reactivated in adult sexuality, at times in forms that we find shameful and unacceptable. From this come the conflicts, whether moral or neurotic, that we all experience in sexual life. (de Lauretis, "Queer Texts," 250–51)

170   **"The subject arrives on the scene rather late in the day"** (Bollas, *Shadow*, 8).

171   **To be in a mood is to broach realms of wordlessness wherein lies our (presumably) truest selves.** Bollas describes his work with three different patients thus: "With all three patients, I was aware that each was preserving something very important—some essential feature of the core of the self—and that this part of the self was recurrently established through moods. . . . Some analysands feel that their moods are the most important authentic memories of their childhood, often because through mood the person feels in contact with a true self-experience" (Bollas, *Shadow*, 113).

171   **Inside our moods, we are put "in contact with that child self who endured and stowed the unrepresentable aspects of life-experience"** (Bollas, *Shadow*, 112).

171   **When he asks, "how far inside the mood is someone?"** (Bollas, *Shadow*, 99).

171   **Here he's following the cue of fellow analyst Paula Heimann** (Bollas, *Shadow*, 1).

179   **"Can you remember when you used to look through these?"** Footage from the 1950s, *The Home Show* on "View-Master: As Seen on TV (50 Years of Video about Our Favorite Stereoscopic Device)," Produced by Eddie Bowers, Mary Ann and Wolfgang Sell. According to the Sells, the episode was called "Fairytales in the Living Room" (Sell, Sell, and Van Pelt, *View-Master*, 95).

177 **Francis could have gone far as the first female Talk Show host.** See Timberg's useful précis of Francis's career ("Arlene Francis").

178 **Florence Thomas is ever out of view.** In a newspaper article written about Thomas at the end of her career, Eleanor Boxx had interestingly summarized Thomas's place in American histories of art: "Florence Thomas is almost unknown to the general public, yet her creations have been viewed by more persons than those of many other American artists" (Boxx, "Sculptor").

Synesthesia for Orphaned Boys

My knowledge and understanding of Charles D. Hubbard's life and work and so many of the guiding factual details herein are indebted to Marius B. Péladeau's biographical essay, *Charles Daniel Hubbard, 1876–1951, American Impressionist: A Biographical Essay*, made to accompany an exhibition of paintings, writings, and drawings at the L. C. Bates Museum, Good Will-Hinckley, Hinckley, Maine, 1996, and published by the museum. Unless otherwise noted, the details of Hubbard's life and work are culled from his meticulously researched essay. This, and Hinckley's ongoing official *Record* of Good Will inform my pages. Throughout the text proper, I use *GWR* to refer to Hinckley's *Good Will Record*, and *CDH* to Marius B. Péladeau's, *Charles Daniel Hubbard, 1876–1951, American Impressionist.*

Ahern, Mary Stokes. "It Really May Be True: '38 Was Only Yesterday." *New Haven Register*, September 29, 1988.

Anderson, Michael. "Francis Lee Jaques and the American Museum of Natural History Bird Halls." Chap. 7 of *Painting Actuality: Diorama Art of James Perry Wilson*. Accessed September 8, 2014. http://peabody.yale.edu/james-perry-wilson/chapter-7-francis-lee-jaques-and-the-bird-halls.

Bunting, W. H. *A Day's Work: A Sampler of Historic Maine Photographs, 1860–1920, Part I.* Gardiner, ME: Tilbury House, 1997.

Chamber of Commerce Journal of Maine. Vol. 17. 1904.

Cultural Landscape Foundation. "Good Will-Hinckley." Accessed June 11, 2015. https://tclf.org/landscapes/good-will-hinckley.

Davis, Dorothy Byrd. "The Closing of Hobgoblin Hall." *Shore Line Times* (New Haven, CT), September 11, 1952.

English, Jean. "Maine Academy of Natural Sciences: An Organic Approach to Education." Maine Organic Farmers and Gardeners Association. Accessed June 1, 2015. http://www.mofga.org/Publications/MaineOrganicFarmerGardener/Spring2012/MaineAcademyofNaturalSciences/tabid/2134/Default.aspx.

"From the Game Fields." *Recreation* 6 (1897): 370.

Gemperlein, Joyce. "Window Dressing in S. Phila. a Long and Personal Tradition."

Philadelphia Inquirer, August 22, 1988. http://articles.philly.com/1988-08-22
/news/26256385_1_window-dressing-hungs-view.

Good Will-Hinckley. "Reverend Hinckley's Vision." Accessed September 24, 2012.
http://www.gwh.org/AboutGWH/History.aspx.

Harvey, Ron, and Nina Roth-Wells. "Collaborative Conservation of the Charles
Hubbard Natural History Dioramas." *Objects Specialty Group Postprints* (American Institute for Conservation of Historic and Artistic Works) 15 (2008):
43–57.

Hinckley, George Walter. "Exhibits at Good Will, Bates Museum." *Good Will Record*
62, no. 3 (March 1949): 47–54.

———. "Exhibits at Good Will, Hubbard Granite House." *Good Will Record* 62, no.
6 (June, 1949): 114–18.

———. "Exhibits at Good Will, Kent Dendrology Building." *Good Will Record* 62,
no. 5 (May 1949): 91–102.

———. *The Good Will Idea*. Hinckley, ME: Good Will Publishing Company, 1929.

———. *The Man of Whom I Write: Incidents in the Life of Reverend George Walter
Hinckley, Founder of the Good Will Homes, Hinckley, Maine*. Fairfield, ME: Galahad Press, 1954.

———. *Sketches at Good Will*. With drawings by Charles D. Hubbard. Hinckley,
ME: Good Will Publishing Company.

Hubbard, Charles D. *Camping in the New England Mountains*. Drawings, text, and
hand lettering by Charles D. Hubbard. Manchester, ME: Falmouth Publishing
House, 1952.

———. "Drawing in the Schools." *Shore Line Times* (Guilford, CT), September 16,
1915.

———. *An Old New England Village*. Drawings, text, and hand lettering by
Charles D. Hubbard. Manchester, ME: Falmouth Publishing House, 1947.

Isenstadt, Sandy. "The Rise and Fall of the Picture Window." In *Housing and Dwelling: Perspectives on Modern Domestic Architecture*, edited by Barbara Miller
Lane, 298–306. New York: Routledge, 2007.

Jetté, Edith Kemper, and Ellerton Marcel Jetté. *American Painters of the Impressionist Period Rediscovered: A Collection Presented to the Colby College Art Museum*.
Waterville, ME: Colby College Press, 1975.

Kanes, Candace. "George W. Hinckley and Needy Boys and Girls." Maine History Online, Maine Memory Network. Accessed July 19, 2014. https://www
.mainememory.net/sitebuilder/site/576/page/933/display.

Little, David. "Charles Hubbard: Wielding a Palette Knife on Katahdin." In *Art of
Katahdin: The Mountain, the Range, the Region*, edited by Carl Little, 91–97.
Rockport, ME: Down East Books, 2013.

Meiburg, Jonathan. "Inside the American Museum of Natural History's Hidden

Masterpiece." *The Appendix*, vol. 1, no. 3 July 2013. Accessed September 1, 2014. http://theappendix.net/issues/2013/7/inside-the-museum-of-natural-historys-hidden-masterpiece.

Noyes, Dorothy. "The Language of Dressed Windows." In *Uses of Tradition: Arts of Italian Americans in Philadelphia*, 66–71. Philadelphia: Samuel S. Fleisher Art Memorial, 1989.

O'Neill, Claire. "In the Background: Art You May Never Notice." National Public Radio, December 14, 2013. Accessed January 4, 2014. http://www.npr.org/2013/12/14/250797697/in-the-background-art-you-may-never-notice.

Péladeau, Marius B. *Charles Daniel Hubbard, 1876–1951, American Impressionist: A Biographical Essay*. Hinckley, ME: L. C. Bates Museum, 1996.

Pray, Leon L. *Taxidermy*. New York: Macmillan, 1947.

Quinn, Stephen Christopher. *Windows on Nature: The Great Habitat Dioramas of the American Museum of Natural History*. New York: Abrams, 2006.

Sarnacki, Aislinn. "Plenty to Draw From: Artists Find a World of Inspiration in the Collections of Curiosities at the L. C. Bates Museum." *Bangor Daily News* (Bangor, ME), August 22, 2011.

Sell, Mary Ann, and Wolfgang Sell. "Remembering Florence Thomas 1906-1991." *Stereo World* 18 (August 1991): 17–19.

Shore Line Times (Guilford CT). "Pastor Pays Tribute to Charles Hubbard." October 4, 1952.

Sturtevant, Lawrence M. *Chronicles of Good Will Home, 1889–1989 at Fairfield, Maine*. Hinckley, ME: Good Will Home Association, 1989.

Sutherland, Amy. "In Hinckley, Lasting Impressions of Maine." *Maine Sunday Telegram*, June 2, 1996.

Vitello, Paul. "Fred Scherer, 98, Dies; Diorama Painter Mastered Even Illusion of Air." *New York Times*, December 8, 2013. Accessed January 4, 2014. http://www.nytimes.com/2013/12/09/nyregion/fred-f-scherer-diorama-painter-who-mastered-even-the-illusion-of-air-dies-at-98.html?_r=0.

Waterville Morning Sentinel (Waterville, ME). "Connecticut Artist Has Painted Over Sixty Oils at Good Will Homes." August 19, 1938.

White, John. "Did You Happen to See—Charles D. Hubbard?" *Times Herald* (Guilford, CT), February 7, 1948.

Wonders, Karen. "Exhibiting Fauna—from Spectacle to Habitat Group." *Curator* 32, no. 2 (1989): 131–56.

———. "The Illusionary Art of Background Painting in Habitat Dioramas." *Curator* 33, no. 2 (1990): 90–118.

———. "The Phantom Vault of Heaven." In *Habitat Dioramas: Illusions of Wilderness in Museums of Natural History*, 192–221. Uppsula: Almqvist and Wiksell, 1993.

182   "An old squirrel has a skin of Herculean strength that makes a satis-
       fying den trophy" (Pray, *Taxidermy*, 73, vii, viii).

185   **Good Will-Hinckley, founded in the late nineteenth century as the
       Good Will Home Association.** Sturtevant's in-house limited-edition
       Chronicles of Good Will Home, 1889–1989 is an indispensable documen-
       tary resource of the history and changes that Hinckley's Good Will
       Homes for Boys and Girls underwent. The institution, as Hinckley con-
       ceived it, would need to have a library, a museum, a school, a chapel,
       and also a working farm. Sturtevant refers to Good Will alternately as
       a "farm home," based on Hinckley's "art of living" (it was sometimes
       called Good Will Home and Farm); a "country home for underprivi-
       leged boys"; and "a culturally enriched pastoral village" (Sturtevant,
       Chronicles, 327). In one of numerous iterations of the project, *The Good
       Will Idea*, Hinckley makes clear that Good Will was conceived neither
       as a reformatory nor as a boarding school nor as a religious institution:
       "The primary object of the Association is to provide homes for those,
       who, through some unfortunate event or circumstance are deprived of
       home life." Though "non-denominational both in teaching and in prac-
       tice," according to Hinckley, religiosity was central to the program, its
       "spirit to be evangelical without being sectarian." Terms of admission
       stipulated no one under nine or over fourteen years of age and that the
       child be "physically fit, of average intelligence, fair morals, and in need
       of a home or a helping hand." Tuition was fifty cents per week "if there is
       someone who can pay. Otherwise at the expense of a benevolent funder"
       (Hinckley, *Good Will Idea*, 55, 61, 83). Hinckley invented a portmanteau
       for his educational philosophy, considered the key to ultimately success-
       ful work for youth, the rather inelegant "Reinsophy," which stood for the
       REligious, INtellectual, SOcial, and PHysical emphases of the home and
       school's itineraries.

186   **The world is out of joint and yet all the world's conjoined, the care-
       fully documented mish-mash seems to say.** What is impossible to tell
       in a contemporary visit to the museum is the extent to which its rooms
       became, of necessity, a repository for a wide array of objects that, de-
       cades prior, had enjoyed more systematic display in designated build-
       ings across the campus, the remains of which are now housed along-
       side and among the original dioramas and collections. A magnificently
       compelling modern ruin or postmortem storehouse, the L. C. Bates
       Museum is tinged with a special beauty and charm that are, in part, the
       effects of its deterioration and disarray. This particular effect was proba-
       bly not part of Hinckley's plan, however synonymous the collections are

with so many centuries-old cabinets of curiosity. If the museum stands now as a repository for lost things, the remnants of a one-time more elaborately choreographed though humbly conceived institution, that's the result of a complex series of events as well as oversights on the part of Hinckley himself, the twists and turns of which are documented by Sturtevant. Following George W. Hinckley's death in 1950 and through to the 1970s when the campus reverted, against his gravest wishes, to a prep school, a great many priceless things went missing, from a group of ancient Chinese vases to a set of fifty contemporary bronze tablets: much was vandalized, outrightly stolen, or, out of negligence, burned to the ground. Hinckley, though an incredibly successful fund-raiser may not have been a very good businessman, and Sturtevant attributes Good Will's fall into chaos to Hinckley's never having built in legal protections for the various endowments he received. He also cites Hinckley's sale of ninety-eight parcels of land including Mount Jefferson in nearby Lee, Maine, bequeathed him by the Jaehne brothers as a major misstep. In light of this history, it is amazing that the museum survived at all.

188   **A room hung with poorly conserved fish carcasses.** As of this writing, and on the day of my third visit to the museum, in early September 2014, Deborah Staber informed me that the museum had just won a large grant designated, in particular, to preserve the fish and a sampling of birds. For other recent conservation efforts, see Roth-Wells et al. ("Collaborative Conservation") who describe the dioramas as "among the oldest and best-preserved . . . in New England." They found Hubbard's background paintings to be "in remarkably good condition," and their deliberations, with the director's guidance, were painstaking as to how best to solve the dioramas' lighting problems (Roth-Wells et al., "Collaborative Conservation," 44). One focus of their repairs was a hyena mount, a great deal of whose fur had been eaten by insects in decades past, other parts of which had suffered deterioration from moisture getting into the case. The fact that taxidermied specimens need to be rendered insect-proof flies ironically in the face of humans' attempts to create death-defying figures from the remains of once-living creatures.

188   **Hunted by Ernest Hemingway himself.** The Hemingway fish came from an anonymous donor from Bimini in 1935.

189   **There was caribou.** A placard in the museum explains that this was one of the last to be shot in Maine and that originally this display held a moose. A photograph in a separate display shows boys at Hinckley looking at the moose diorama, the moose having been replaced by the caribou sometime between the 1940s and the present.

190 **The center of a diorama sports a 3-D squirrel, or perhaps it's painted in?** This particular diorama also features a taxidermied chipmunk, Eastern gray squirrel, red squirrel, long-tailed weasel, and short-tailed ermine.

191 **Apparently forgotten American impressionist.** A ninth-generation New Englander, descended from the founders of Guilford, Connecticut, where he spent most of his life—he was born in Newark and moved to Guilford at age four—with no apparent familial influence in his determining to become an artist, Hubbard decided in his late teens to pursue artistic training first and intermittently, as money allowed, at Yale University's School of Fine Arts and, later, at New York City's Art Students League. Among his teachers were the well-known muralist Kenyon Cox (1856–1919) and the American impressionist painter John Henry Twachtman (1853–1902). In his beautiful biographical essay on Hubbard's life's work, former Farnsworth Museum director Marius B. Péladeau discovers an unassuming but vivacious man who was a beloved instructor of art at Hopkins Grammar School in New Haven and, later, at the then Commercial High School where he became head of the Art Department, "commuting to the city each day on the trolley from Guilford" until his retirement at the age of seventy-two in 1948 (*CDH*, 11). Hubbard was hardly an obscure artist, according to Péladeau: he actively exhibited his work and was a member of the Salmagundi Club; he showed at the National Academy of Design; and he enjoyed a steady stream of commissions not only for his illustrations but occasionally also for portraits in oil and for murals. Péladeau believes that word of his work at the L. C. Bates Museum must have spread southward since, in 1925, he was hired by the Yale Peabody Museum of Natural History to produce at least two murals for exhibits that opened that December. Those particular creations, like a great deal of Hubbard's work, and unlike those in the L. C. Bates Museum, however, disappeared.

192 **A stuffed (double-wattled) cassowary.** This depends on where the cassowary is placed on the day of a visit. On this particular day, it was on floor level; on other visits to the museum, I noticed that it was set high above the bird dioramas (L. C. Bates Museum).

195 **The school logo, in the shape of a medallion or "roundel."** Though Hubbard designed the ornamental seal for the school, site unseen, Hinckley was taken with it from the start as a perfect representation of his ideals. The original drawing from 1912 appeared on the masthead of the *Good Will Record*, on watch charms, on tokens, on postcards, and on stationary. In 1917, Hubbard was asked to execute a full-color

oil-painting version of the seal that could be mounted in the newly appointed Prescott Hall. For the new version, he found a model in a sixteen-year-old Guilford boy, Merle Jillson. Many decades later, Jillson's widow would report how "Merle could never overcome a certain feeling of inadequacy in fulfilling the ideal role in which he was cast by Hubbard in their sessions in the winter of 1918" (Sturtevant, *Chronicles*, 101). As far as the record shows, aside from this early medallion commissioned by Hinckley, and though Péladeau refers to him having received some payments "out of the Bates Gift" (Péladeau, *Charles*, 8) for the museum renovation, most of the prodigious painting, creating, building, and designing that Hubbard did at Good Will, he did without remuneration (Sturtevant, *Chronicles*, 102). He had copied into his common book—a daily journal he kept in careful calligraphy starting in 1917—a verse composed by Kenyon Cox as guiding principle: "Work thou for pleasure; paint or sing or carve / The thing thou lovest though the body starve. / Who works for glory misses oft the goal, / Who worries for glory coins his very soul. / Work for work's sake, then end it well; maybe / That these things shall be added unto thee." Hubbard lived this dictum, quite literally. Of commercialism, he wrote: "I never liked commercialism, and hated the city. And it was partly for these reasons that I largely gave up commercial art and took up school teaching" ("Did You Happen to See").

195–96 **A plow, a machinist's hammer, a baseball bat.** As explained in a Maine Memory Network posting, "The painting includes a tree to represent forestry, a plough for agriculture, a machinist's hammer for mechanical interests, a baseball bat for organized athletics, and books for science and literature" (Kanes, "George").

196 **But he and Hubbard become collaborators in the mutual creation of roomlets, nooks, interiors, and interiorities: of birthing grounds for moods.** Though my emphasis is on the unusual partnership that Hubbard and Hinckley formed in the cocreation of a kind of Good Will aesthetic, it bears mentioning that Good Will Hinckley was a complex enterprise run by a great many people. Some of the early planning of its parklike features, winding village roads, and pond enjoyed the direct influence of Charles Rust Parker of the Olmsted Brothers Firm whose designs were implemented as early as 1914 (Cultural Landscape Foundation, "Good-Will"). Sturtevant notes Hinckley's assembling an advisory committee of experts in executing the larger museum plan: "William M. Swain of Farmington in ornithology; David L. Dorwood, a former Good Will boy and graduate of Yale School of Forestry; T.A. James of the Yale

Museum; John Francis Sprague of Dover-Foxcroft in history; Loren
Merrill of South Paris in minerology; Mrs. Virgin DeCoster, the Maine
'Butterfly Woman'" (Sturtevant, *Chronicles*, 203). Hinckley forged part-
nerships continuously with numerous people, especially benefactors to
the home and farm. One story of many has him meeting Edwin Bancroft
Foote in New Haven. When the scheduled speaker, Jacob Riis, fell ill,
Hinckley had to replace him, giving the dedicatory address to the Edwin B. Foote Boys' Clubhouse there. Later, Hinckley visited Foote in a
winter cottage in Rangeley, Maine, where he read to him testimonials
from Good Will alums. Sturtevant describes the soon-to-be-benefactor
Foote as a scrooge-like figure who softened before the letters from children at Good Will (Sturtevant, *Chronicles*, 70). Hinckley had attracted
"a highly competent and loyal farm and maintenance staff" to the place
who, though "they didn't get paid much . . . lived well"—they were Walter and Reginald Price; Marshall and Raymond Gifford; and Roy Fenla-
son. These farmers "worked 12 and more hours a day as a steady thing,
day after day, year after year" (Sturtevant, *Chronicles*, 163). They were
aided, crucially, by the children at Good Will who engaged in everything from ice harvesting (before the age of freezers) to wood piling,
gardening, bringing the cows to pasture, and caring for the poultry. The
sight of twenty-five to thirty boys chopping wood in snow was a scene
that Hinckley loved: in two weeks, Good Will boys would cut and pile
one hundred cords. Insofar as Hinckley was, according to Sturtevant,
himself a (frustrated) artist, Hubbard brought the artistic vision to the
campus that Hinckley longed both to fulfill in himself and to bestow
on the campus: Hinckley's self-description in his *Record* and elsewhere
depicts "a lover of art . . . [who], had religion not got the first place in
his soul . . . would have devoted his life to palette and brush . . . to become an artist was his first cherished ambition" (Sturtevant, *Chronicles*,
265). Sturtevant can't seem to decide who the real artist of Good Will
was: Hinckley or Hubbard. "Through Hubbard, the artistic disposition
of Hinckley found expression," he writes. Then: "George Walter was the
crown, But Hubbard was the jewel in the crown." Or again: "George
Walter was the artist of Good Will campus and Charles D. Hubbard was
his assistant" (Sturtevant, *Chronicles*, 102; 204).

197 **He insisted, instead, on building small cottages for the boys and
girls.** Good Will cottages generally held no more than fifteen girls or
boys and a matron.

198 **Shell House included four murals by Hubbard of Maine seacoast
scenes.** According to a 1938 *Waterville Morning Sentinel* article, Hubbard

was "most satisfied" with the paintings he made for "Shell House, Rock House, and the House of Dendrology": "These are all murals, not backgrounds. The walls of the Shell House are decorated with four scenes from the Maine seacoast. 'Low Tide at Owl's Head, Bar Harbor,' 'Looking from Sorrento,' 'The Ovens at Bar Harbor,' and 'On Mount Desert'" (*Waterville Morning Sentinel*). These latter are among the works that went missing in the period of Good Will Hinckley's reformation and decline: "The shell house paintings are not on campus," Deborah Staber informs me, "I do not know what happened to them" (e-mail correspondence, July 26, 2014). The building itself was moved in 1968 to near the current high school and "used for many things including a vegetable stand. . . . It does still have two little shells that Hubbard carved on the doorway, but someone painted them gold" (e-mail correspondence, July 30, 2014). Shell house was based upon gifts from the Jaehne Brothers and arranged under the scientific direction of Florence A. Nelson. "Granite house is still on the trail but in poor condition with no doors or windows." Of the twelve murals made for Granite House—four large and eight small— the former are in storage at the L. C. Bates Museum while the latter are lost. They represented the Maine quarries that Hubbard and Hinckley had traveled to in search of specimens, including: Mount Kineo, Limerick, Tenants' Harbor, Topsham, Monson, Mount Waldo in Frankford, South Addison, North Jay, Rockland, Norridgewock, and Waterville (*Waterville Morning Sentinel*). At one time Granite House also featured a "garden of rocks" planned by Hubbard and described in one of his obituaries as "one of the most attractive places at the Good Will Home born of his vision" (*Shore Line Times*). Hubbard and Hinckley apparently also gave serious thought to the aesthetic of these buildings' interiors, writing of Shell House that "all the woodwork in the building is stained dark brown in order to bring out the form and color of the shells by contrast," and of Granite House that "the outside of the logs are oiled . . . the inside is finished in Venetian red, and pearl gray" (Sturtevant, *Chronicles*, 206, 205).

198 **One of an apple tree in spring.** Péladeau cites Hinckley's *The Man of Whom I Write* for even more exacting data about the trees, including the fact that the oak tree was supposed to be the particular oak "on Nut Plain in Guilford under which Lyman Beecher courted Roxana Foote" (Péladeau, *Charles*, 36n73).

199 **The magisterially proportioned and sleekly realistic dioramas that would be the standard bearers of their day.** Stephen Quinn describes New York's American Museum of Natural History as the "Louvre of di-

orama art." These were, for the most part, dioramas on a grand scale that give the effect of "nature . . . enshrined in a grand architectural space," arriving, in the 1930s, in the famed Akeley Hall with displays that were twenty-three feet deep, with eighteen-foot ceilings and window panes that were thirteen feet high (Quinn, *Windows*, 18). The only habitat dioramas in New England older than those in the L. C. Bates are to be found at the Fairbanks Museum and Planetarium, Saint Johnsbury, Vermont, created between 1890 and 1915 (*CDH*, 33).

199 **Hubbard isn't a Carl Akeley or Teddy Roosevelt strangling leopards in the wild.** "I have never had any exciting experiences with wild animals. I have never been attacked by bobcat nor a bear, not having been a mighty hunter, but only a peace-loving artist," Hubbard wrote (Hubbard, *Camping*, 67). In a bio-sketch for the *Times Herald*, he seemed to take pleasure in describing his life as "tame": "It can hardly be otherwise, as I have never been shipwrecked, nor appeared upon the field of battle, nor divorced, and thus far I have not even been sent to jail" ("Did You Happen to See").

203 **But the habitat diorama, gaining its first foothold in particular in Sweden and the United States.** Karen Wonders's fabulous book and articles on the history of the habitat diorama provide the context for my understanding of the form and, consequently, of Hubbard's singularity.

204 **"Artists to document, record"** . . . **"most imperiled"** (Quinn, *Windows*, 6, 79). Though Chapman's name as leading ornithologist, curator, and designer is the one most associated with the displays in the Hall of North American Birds, Quinn names the museum's first team of background artists dating to the turn of the twentieth century: "Hobart Nichols, Charles Hittel, Carl Ringius, Robert Bruce Horsfall, and the renowned bird painter, Louis Agassiz Fuertes." I would also add Cara Horsfall, who appears alongside her husband in an adjacent photograph (Quinn, *Windows*, 80). Though female taxidermists, dioramists, and background painters are few and far between in the written history, Michael Anderson draws our attention to Boulder-based naturalist and taxidermist Martha Maxwell's displays for the 1876 Centennial Exhibition as compelling precursors to the designs Chapman would conceive soon after for the American Museum of Natural History, where he was hired in 1888. On the wondrous history of the Whitney Hall of Oceanic Birds—the brainchild of Chapman and whose dioramas were painted by Francis Lee Jaques—and on its once magical and now dismantled and obscured place in the American Museum of Natural History, see Meiburg ("Inside").

204   **"The aimless visitor involuntarily pauses"** (Wonders, "Phantom Vault," 130).

204   **Either a Vermeer-like devotion or a brief van Gogh passion.** Background painter, William Traher: "It takes a Vermeer-like devotion rather than brief Van Gogh passion to stay wedded to one painting as long as a diorama artist must" (Wonders, "Illusionary," 111).

204   **"None of the early American diorama painters had an artistic training whereby they could apply the latest knowledge of Impressionism to their particular genre"** (Wonders, "Phantom Vault,"199). Anderson makes the interesting point that, though impressionism was outrightly "denigrated" in the creation of natural history displays, it still may have "informed many a diorama painter's work"—a kind of impressionist method in the service of realism, explaining that "Steve Quinn instructs his art students at the American Museum to view the background paintings through binoculars. They are surprised to see just how loosely the paintings are handled and how much high-keyed color was used" (Anderson, "Francis Lee Jaques"). Quinn references Fred Scherer's more pointillist paintings done for the reopening of the North American Bird Hall in the early 1960s as further evidence of the persistence of a realist convention: "It may come as a bit of a shock to find such a subjective and highly personal interpretation of nature in a science diorama, but the technique is so well executed that the paintings do not interfere with the educational goals of the exhibits, and they enjoy a wide and loyal following" (Quinn, *Windows*, 80).

205   **Carl Ringius's painting remained unfinished** (Quinn, *Windows*, 100).

205   **"Quiet tints that don't attract away from the specimens."** Wonders quoting William Hornaday, "father of modern taxidermy," who, like Carl Akeley after him, began his career in the late nineteenth century with the main museum supplier of diorama specimens at the time, Rochester-based Henry A. Ward's Scientific Establishment (Wonders, "Phantom Vault," 126).

205   **Not a "window onto nature" . . . not intent on making us feel that we are "standing before an actual scene"** (Wonders, "Phantom Vault," 205). Hubbard describes the nonrealist principals that inform his aesthetic in an article composed by Dorothy Byrd Davis on the closing of his studio just prior to his death: "Most people go right up to a painting, but that is like not seeing the forest for the trees. You have to move away a little from it, and never tell where the picture was painted. Any camera will do that. A picture should tell a story, and the artist ad-

justs the subject to get that across. If the painting is a literal study, it fails. His message should evoke a response" (Davis, "Closing," 81). Surprisingly, according to one report, Hubbard was never entirely happy with his diorama work. "Although these painted backgrounds form the largest group of his creation and have been enjoyed by thousands of visitors to Good Will, artist Hubbard has never been completely satisfied with them. To make an appealing exhibit, he believes the carpenter, the taxidermist, the background painter, and the electricians need to cooperate closely. Unfortunately, the limits of time, space, and money usually make this impossible. The exhibits fall short of perfection even at the Peabody Museum of Yale University when Hubbard painted some backgrounds, all conditions were not satisfactory. Mr. Hubbard looks forward to the time when the Bates Museum at Good Will can afford a curator to attend to such details" (*Waterville Morning Sentinel*).

206 **We should never be inclined to look down at an eagle in mid-flight.** Perhaps Hubbard was inclined to agree: "The eagle in the Hubbard Diorama was originally higher on the tree—the tree broke over time and when we worked on the diorama it was felt best to leave it as it was. In old photos you can see it was placed for the eagle to have a better view and seem to be high up in the . . . tree" (Deborah Staber, e-mail correspondence, December 3, 2014).

209 **Hinckley composed a book-length account of his life.** Hinckley wrote two other autobiographies in addition to this one: *Chapters of the Record* (1914) and *As I Remember It* (1943).

209 **It's not just his curiously inexplicable German accent.** The hard-to-place accent is apparent in film footage of Hinckley currently on view at the L. C. Bates Museum. If Hinckley was without charisma, he wasn't without the ability to persuade, enjoin and convert. Hinckley originally came to Maine to represent the American Sunday School Union of Philadelphia. Beginning in Hermon, outside of Bangor, he developed a reputation as an itinerant preacher, organizing thirty-four Sunday schools in rural Maine while based on Newbury Street in Bangor. During this time, he preached in churches, sang and talked in schoolhouses, entertained in homes, hunted, climbed, and fished. The Good Will Home Association was formed on October 5, 1889. Sturtevant marks its heyday from 1889 to 1915; its middle age from1916 to 1941; and its decline following Hinckley's death in 1954. The first children at the school, twelve girls and three boys, arrived in the summer of 1889. The girls were sent from the Rutland Street Girls Home, and the boys came from

Dedham Home in Massachusetts, founded by George Henry Quincy who was also one of Hinckley's first benefactors. At the end of the summer, the girls returned, but the boys stayed, thus constituting the first of Hinckley's charges: Joseph Jordan, Harry Jordan, and Mason Parker. Fifty years later, these boys returned to Camp Quincy for an emotional celebration. The history of Good Will's benefactors forms a fascinating substrate to the whole. In addition to those Hinckley convinced, far and wide, to support the project, he tells tales of children themselves supplying funds for the school as in the case of the boy who donated the nickels he collected each day for not breaking while washing dishes to start a cottage for girls (Hinckley, *Good Will Idea*, 24–26). Sturtevant cites one of Hinckley's sons as saying that his father was "always opposed to the use of the family name in connection with the school" (Sturtevant, *Chronicles*, 320), but it seems fairly evident that the name of the school echoed with its founders initials: "GW."

210 **"He was slender, boyish, aggressive; he had red hair, blue eyes, freckles."** Sturtevant quotes Hinckley in sentences that make Mason read as Hinckley's double: "On the sixth of March, 1869, I stood in the aisle of the First Church (Congregational) and entered into the covenant with the church; the 'other fellow'—the boy I had prayed for when I offered my first real prayer for myself . . . stood by my side." This boy was Ben Mason (Sturtevant, *Chronicles*, 6).

212 **Ben Mason comes to signify the "class of boys to whose welfare [Hinckley] was to devote [his] life."** Hinckley refers in his records to his particular interest in helping "sons—especially youthful sons— boys" (*GWR*, March 1949). In addition to farming, camping was an integral component of the Hinckley school's ideal. Some would even refer to Hinckley as the father of the boys' camp movement, whereby "camping" and "camp" aren't only understood as leisure activities but as adjuncts to physical, intellectual, and moral development. Like Summer Dudley, founder of the YMCA, Hinckley thought of camp as a better means to minister than the pulpit. Prior to founding his home, he tested his theory by persuading parishioners in West Hartford, Connecticut, to "surrender their boys to my care for a couple of weeks." Once he established Good Will, he built in one- to two-week-long camping trips on Sebasticook Lake for the students at the end of the farm season. (During one such trip, two cottages burned to the ground with no apparent cause. That year, all of the winter clothes were destroyed in the fire.) Hinckley regularly went on camping trips to recover from the stresses incurred

in running Good Will. Usually, he would bring a boy with him (or more precisely) an alumnus of Good Will, as in 1904 when a need for "prolonged rest" took him into woods in the Moosehead Lake region for a stay of several weeks. "Henry Blake, a former Good Will boy, now a student at Dartmouth College, will accompany [me] on [my] trip into the woods," he writes. "I am able to be out but I am not able to be in—in the office," Hinckley wrote. "We will talk some. . . . There will be no one to talk to but Henry; no one to crowd and jostle" (Sturtevant, *Chronicles*, 65).

212–13 **A painting of Ben could . . . function in the way the Good Will's branding roundels did—as a sort of advertising seal of the home's mission.** A kind of coat of arms sporting the school's motto, "He Works When He Works—He Plays When He Plays—He Is Strong on Individual Effort—Yet He Labors for Community Good," the 1924 roundel pictures a busy beaver at its center, and a placard explains that the beaver was chosen as a symbol by the children themselves, to represent the character of Good Will boys and girls.

214 **Portrait in oils.** The small number of portraits Hubbard was known to have painted in the course of his career seem mostly of Good Will personages, e.g., Hinckley himself as well as his sister, Jane E. Hinckley; his son, Walter Palmer Hinckley; his wife, Harriet Palmer "Mother" Hinckley; and, Moses Giddings, the organization's first president. Péladeau also notes a commissioned portrait of Dr. David N. Beach dating to 1922 that still hangs in the Bangor Theological Seminary library (Péladeau, *Charles*, 8).

215 **"Do not haggle the skin at its margin under the antler burrs"** (Pray, *Taxidermy*, 7).

215 **Gifford's business card** (Bunting, *Day's Work*, 220).

215 **"The first boy to make a great response to Hinckley's small museum"** (Sturtevant, *Chronicles*, 201).

215–16 **A mounted loon that resided there, "captured and mounted by Reeves"** (Sturtevant, *Chronicles*, 202).

216 **His teacher . . . Mr. Gifford, had mounted the infamous caribou.** In the March 1949 *Record*, Hinckley explains that the caribou, mounted by Gifford, was donated in 1931 by Mrs. Marjorie McGuire, a Good Will High School graduate. He goes on to remind us that Gifford had instructed Walter Reeves and that Reeves "maintained his interest [in] and added specimens to the Good Will museum" (Hinckley, *Good Will Record*, 54). Drawing from Hubbard's commonplace books for 1922 and

from a newspaper article dating to that time, Péladeau, referring to the mammal room, writes that "most of the taxidermy was done by Walter Reeves" (Péladeau, *Charles*, 24n36).

216–17 **When Gifford instructs him "to lay the animal on its back"** ("From the Game Fields," 370).

217 **Is Walter meant to become like Spinoza, who, again, cited by Hinckley cited by Northrup cited by Froude.** The world that boys and girls at Good Will were ushered into was in many ways its own culture, full of ceremony and ritual. Among the numerous symbolic totems—from flags and cups to medallions—features a "Good Will Emblem" made from six pieces of symbolically chosen wood, underneath which lies a silver plate engraved with "one word in Old English characters" that, each year, was "bequeathed in silence and accepted in silence" to the junior class by the senior class (Hinckley, *Good Will Idea*, 63). Hinckley maintained order at the school with shaming-type punishments, like having a student's name posted for three months in one of the home's public halls, or the seldom-imposed sequestration in an attic for several days with nothing to eat but crackers and water. (Sturtevant, *Chronicles*, 106). Tales of love and of death would come to mark the community, from the story of the matron of a cottage, Myra Porter, marrying a cottage boy to the tragedy of a boy who drowned in the nearby river when venturing on ice after being told not to. A very special boy to Hinckley and he who handled the stereopticon for Hinckley's lectures, Daniel Alexander McDonald, died of undetermined causes in 1903. Hinckley memorialized him in his tribute, "The Story of Dan McDonald, 1885–1903." (Sturtevant, *Chronicles*, 59). Particular teachers became known as perfect maternal substitutes, e.g., Hinckley's sister, Jane E. Hinckley, a self-educated matron of Good Will Cottage (1889–1914) made lasting impressions on the boys, who described her as "saintly" and the "picture of [the] mother they left behind," before she, too, left them behind, dying young from stomach cancer. A reassuring teacher and friend, an independent woman, she founded the boys' choir and was known for being a discerning bibliographer who built up the library. (Sturtevant, *Chronicles*, 98–99). At one of the dedications to a cottage in 1909, a professor from Bates College noted the anti-institutional effects of life and learning at Good Will: "The boys before me are not dressed alike, do not sit alike—their singing is unlike the singing heard in our public schools. The individual tone rings out in the volume of sound" (Sturtevant, *Chronicles*, 72). Some children stayed at Good Will for five to ten years; some stayed on and worked there; others went on to work

on farms, shops, mills, and white-collar jobs in various cities. Among its distinguished alumni, the school counts a president of the University of Hawaii (Sturtevant, *Chronicles*, 56).

217   **Since he and a brother maintained a taxidermy practice in Portland** (*Chamber of Commerce Journal of Maine*, vol. 17).

223   **Into this seasonal diorama, I would pose and stow plastic figurines of animals that sometimes got lost in the paper hay.** Only retrospectively have I come to consider the characteristically tripartite windows of the row homes peculiar to South Philadelphia where my Sicilian grandparents lived and where my father grew up as precursors in a mood lexicon of dioramas, though I cannot know if they captivated my father as they did me. Similar to the arrangements on tombstones in Sicilian cemeteries—a photo here, a plastic flower there—the windows are like fronts to a vault that holds a dead body: the house as mausoleum. Tabernacles of remnants of a half-remembered "old country" where lace or beaded curtains fronted the sea, these windows are offset arrangements of individualizing shapes and mix-and-match materials—ceramic, plastic, paper—a Madonna, a hobgoblin, a garland—constituting combination mass-produced and handmade kitsch. As a child, I experienced them as rows of carnival booths laced with wonder and surprise. A horizontal window made of one central large panel braced by two slender panes of glass, the South Philly architectural feature wants to be called a "picture window" even though the only thing it has in common with the midcentury American middle-class suburban type, or "landscape window," is its shape. Folklorist Dorothy Noyes appears to be the only scholar to date to have taken an interest in what she calls "the dressed window." Noyes places the (often secular) vernacular assemblages in the context of women's votive traditions and home altars of Italian Catholicism as well as the niches and outdoor shrines on house fronts that are common in Naples. The windows communicate everything from Italian Americans' reinterpretation of popular culture to the state of health or well-being of the family inside the house (Noyes, "Language"). See Isenstadt for a fascinating overview of this architectural form and its place in a larger politics of class ("Rise").

226   **Her personal favorite specimen.** My own favorite item in the museum is rural Mainer, amateur entomologist Mattie Wadsworth's (1862–1943) butterfly collection, housed in the thirty-drawer wooden cabinet she made for it. Other of her collections reside in the Pennsylvania Academy of Science and the Smithsonian Institution. A dragonfly is named for her.

226   **The correspondence between one Walter Smith and G. W. Hinckley.**
According to Sturtevant, Smith gave a large number of mounted birds to
the museum, though his first gift to Good Will was seventy-five pairs of
blankets. Maine-born Walter M. Smith, of Smith, Hogg and Company,
Stamford, Connecticut, woolen goods merchants, became a major benefactor of the Pines Assemblies of 1897–1906 at Good Will Hinckley,
a kind of Chautauqua featuring "travelers, naturalists, geologists, and
boys' writers" who would "gather and address not only the Good Will
Boys, but those who had come from afar." As a prime promoter of the assemblies in the pines, Smith built his own cottage in the woods, dubbed
Camp Rest Awhile. The assemblies came to a close in 1906 following
Smith's declining health and Hinckley's 1904 nervous breakdown, but
Camp Rest Awhile survived until 1949 when it was burned in a fire.
Even when the assemblies overran income, Smith continued to subsi-
dize them. His links to philanthropists like Thomas Hall, A. P. Ryerson,
and Andrew Carnegie made possible the formation of a girls' farm in
1905, as well as a new library thanks to a $15,000 endowment with a
stipulated in-kind gift from Carnegie. The Elizabeth Wilcox Smith Cottage, considered the best-built cottage on the campus, was named for
Smith's deceased daughter (Sturtevant, *Chronicles*, 31-34; 76).

228–29 **The air is purer here.** Charles Hubbard was painting air, there's no
doubt about it: in that, he was, like other background painters, attempting to find a form for the imperceptible. Rather than create a visual illusion of air as space, however, he created enclosures whose effect was to
enable the person who stood before them—which is also to say, inside
of them—to breathe. See Scherer, the well-known diorama painter's
comments on the challenges, with dioramas in the grand tradition, of
creating an illusion of "depth and substance" in the "airspace of the picture" (Vitello, "Fred Scherer").

229   **The newly conceived high school that now resides there.** Under
the direction of Emanuel Pariser and Troy Frost, the Good Will campus is currently experiencing a rejuvenation as the newly formed (2011)
Maine Academy of Natural Sciences that aims to "to offer to high school
students hands-on, project-based learning focusing on agriculture, sus-
tainability, forestry, business, alternative energy and the environment."
According to an article about the school on the website of the Maine
Organic Farmers and Gardeners Association, the academy draws its
funding from a variety of sources, including the state, grants, some parents' tuitions, school districts that support students wanting to attend
the school, and "scholarships from the former school's endowment"

(English, "Maine"). On Good Will-Hinckley's website, the new school's mission is described as in keeping with the legacy of Good Will.

Picture Books

Dauenhauer, Bernard P. "The Phenomenon of Silence—First Approximations." In *Silence: The Phenomenon and Its Ontological Significance*, 3–25. Bloomington: Indiana University Press, 1980.

Brown, Margaret Wise. *The Country Noisy Book*. Illustrated by Leonard Weisgard. New York: W. R. Scott, 1940.

———. *The Indoor Noisy Book*. Illustrated by Leonard Weisgard. New York: W. R. Scott, 1942.

———. *The Noisy Book*. Illustrated by Leonard Weisgard. New York: Harper and Row, 1939.

———. *The Quiet Noisy Book*. Illustrated by Leonard Weisgard. New York: Harper and Brothers, 1950.

———. *The Seashore Noisy Book*. Illustrated by Leonard Weisgard. New York: W.R. Scott, 1941.

———. *The Summer Noisy Book*. Illustrated by Leonard Weisgard. New York: Harper and Brothers, 1951.

———. *The Winter Noisy Book*. Illustrated by Charles B. Shaw. New York: Harper and Row, 1947.

Laartz, Paul. "Photo-Finish Was Only Beginning for Portland Concern." *Sunday Oregonian* (Portland, OR), May 26, 1948.

Lecourt, É. "The Musical Envelope." In *Psychic Envelopes*, edited by Didier Anzieu, 211–35. London: Karnac Books, 1990.

Marcus, Leonard S. *Margaret Wise Brown: Awakened by the Moon*. New York: Harper Collins, 2001.

Oregonian (Portland, OR). "Girl Wins Fellowship: Florence Thomas of Portland to Study Sculpture in Europe." June 11, 1931.

Oregonian (Portland, OR). "Work of First-Year Class at Museum of Art Shown." December 31, 1933.

Praeger, Joshua Harris. "'Goodnight Moon' Couldn't Protect Heir from Life's Nightmares." *Wall Street Journal*, September 8, 2000. http://www.wsj.com/articles/SB968365456431553434.

Schwartz, Hillel. "Quiet Noisy Books." In *Making Noise: From Babel to the Big Bang and Beyond*, 768–94. New York: Zone Books, 2011.

Sell, Wolfgang, and Mary Ann Sell. "Remembering Florence Thomas 1906–1991." *Stereo World* 18 (July–August 1991): 17–19.

Smith, Virgil S. "Fairy Tales Pictured in 3 Dimensions for Exhibiting through Stereoscopes." *Oregonian* (Portland, OR), February 9, 1949.

234   **There are books which are the result of research.** A detailed description of the content and aims of the two murals written, without a doubt, by Hubbard himself, appears in a pamphlet at one time distributed by the library.

234–35   **By her having sculpted in a mane for the rocking horse.** The Sells explain that "Thomas began developing her sculpting talents while working in her father's ornamental plaster-casing and cast-stone firm" (Sell and Sell, "Remembering Florence Thomas," 17). She studied at the Art Institute of Chicago and at University of Oregon under the sculptor Avard Fairbanks and won a fellowship to study in Europe for two years in Munich and Vienna (*Oregonian*, 1931; 1933). In addition to the Alice in Wonderland plaque, she produced architectural sculptures, having "designed and sculpted the original models for the newel posts at Timberline Lodge on Oregon's Mt. Hood. Wood carvers on that 1930s WPA project then used her models to create the finished animal figures we can see and touch today." These may be the only extant sculptures from Thomas's hand. The Sells write, "It is sad to know that most of the wonderful work done by Miss Thomas has been destroyed over the years" (Sell and Sell, "Remembering Florence Thomas," 18, 19). When asked about the fate of the body of work that constituted her View-Master tabletop designs, Wolfgang Sell replied: "[View-Master] Plant needed room so figures went out 2nd floor window into a dumpster. We have a group of those that were saved. Have not seen any photos by her or know of any other art work. But we have never looked" (e-mail correspondence, May 27, 2015).

240   **Brown had named the child of a friend the right to all monies earned by her books** (Praeger, "Goodnight").

241   **But Brown made sure the larder was stocked with "champagne, fresh cream, imported cheeses and other such necessities"** (Marcus, *Margaret*, 163).

241   **The mirrors were hung across from a door that opened onto a sheer fifty foot drop.** This, and all previous references in this paragraph are to Marcus (*Margaret*, 165).

242   **"I used to bring live things from the garden."** (Boxx, "Sculptor"). An earlier newspaper article (dating to 1949) provides additional illuminating details of Thomas's practice: that her View-Master tabletop figures generally did not exceed six inches in height and that they were painted with oil colors and nail polish. Because the clay figures don't bend, Thomas needed to make separate figures for each new posture. Some of the materials used to build the settings included "moss, paper, asbestos,

soil, stones, hair. She obtained a realistic lake with a piece of Plexiglas with ripples of transparent household cement" (Smith, "Fairy Tales").

242   **She continued to make 3-D constructions, but now by way of what was known as the *Personal* View-Master camera.** See Sell and Sell: "Florence retired in 1971. Upon her retirement, she spent more time communicating with nature working in her garden on a daily basis. Her beloved cat would spend hours watching her while she worked out abstractions in wood in her daylight basement studio. Another favorite pastime was taking pictures with her View-Master *Personal* Camera. She continued to live in Portland until her death in 1991" (Sell, Sell, and Van Pelt, *View-Master*, 97).

242   **He lived at an apparent distance from his wife.** Hubbard was married to Edith Leland, a fellow artist whom he'd met at Yale School of Fine Arts, though you wouldn't know it since it seems he regularly left his wife behind on his prolonged and frequent solo treks to Maine. The only trace of her in Marius Péladeau's account is that she was "sickly most of her life" and that "there were no children of their union" (*CDH*, 31). Scant traces of Edith's life in the Guilford Library holdings show her keeping a journal in 1920 with numerous entries marked, "Charlie away"; the brief daily notations are not without observations of beauty or the occasional humorous aside. Numerous of Hubbard's journal entries read, "Edith doing poorly."

243   **Perhaps Brown thought designing a "Picture Window"** (Marcus, *Margaret*, 236).

243   **The Noisy books, Brown explained, "came right from the children themselves"** (Schwartz, "Quiet Noisy," 790).

244   **"Dogs," Brown wrote, "will also be interested in 'the Noisy Books'"** (Schwartz, "Quiet Noisy," 790).

244   **For isn't it the nature of sound to emanate from the object world but remain detached from it.** Lecourt, in "The Musical Envelope," summarizes some defining conundrums of sound thus:

> The problematic of sonority is characterized by a certain number of parameters, of which I shall only mention the principal ones: first of all, *the absence of boundaries*. Absence of boundaries in space: sound reaches us from everywhere, it surrounds us, goes through us, and in addition to our voluntary sonorous productions, sounds even escape surreptitiously from our own bodies. Absence of boundaries in time: there is no respite for sonorous perception, which is active day and night and only stops with death or total deafness. Sonority is also characterized by *the lack of concreteness*. Sound can never be

grasped; only its sonorous source can be identified (not always), modulated or even manufactured. The sound-object is an acoustician's construct, not a fact of experience. Finally, sonorous experience is one of *omnipresent simultaneity* (211).

245 **Philosophers of silence spoke of silence as its own active presence, not merely a negation of sound.** I am paraphrasing a number of the premises of phenomenologist Bernard P. Dauenhauer's magnificent book *Silence: The Phenomenon and Its Ontological Significance*, and in particular the chapter titled, "The Phenomenon of Silence—First Approximations."

249 **This time an abstract painter other than Weisgard, Charles G. Shaw.** Among other things that Shaw, a life-long confidante of Brown, came to be known for, he was the author of a popular children's book that Marcus describes as "a clever counterpart to Margaret's Noisy series," *It Looked Like Spilt Milk* (Marcus, *Margaret*, 217).

252 **"For what affects us in the act of reading involves the present of the past in substance—not the sign of the past or its representation"** (Gumbrecht, *Atmosphere*, 14).

256 **"Although he was 63 years old in 1938"** (Ahern, "It Really May").

257 **Each pause, a fore- or after-silence that punctuates.** Fore- and after-silences are Dauenhauer's coinages and refer to the silence that immediately precedes an utterance and the silence that terminates an utterance (Dauenhauer, "Phenomenon").

259 **That letters are something we learn to draw.** I once heard the cartoonist Lynda Barry spell out a thesis that was as straightforward as it was startling: that there is no rhyme or reason to our being trained away from picture books and toward the presumably higher purpose of words, words, words; that we suffer as child readers a staunching of the imagination and of individual and collective growth by not being taught to draw over the same course of the years that we are taught to read and write.

Mood Telephony

275 **The "auditory apparatus," and the "phonatory apparatus"** (Nancy, *Listening*, 29).

The Tic-Tic-Tic of a Dime Hitting the Floor

Prevenas, Nick. "UA Cancer Center Surgeon Assists Patients Suffering from Chemotherapy-Related Hearing Loss." University of Arizona Cancer Center. Updated February 26, 2013. http://surgery.arizona.edu/in-the-news/ua

-cancer-center-surgeon-assists-patients-suffering-chemotherapy-related
-hearing-loss.

The Exciting or Opiatic Effect of Certain Words

Kristeva, Julia. "Life and Death in Speech." In *Black Sun: Depression and Melancholia,* translated by Leon S. Roudiez, 31–68. New York: Columbia University Press, 1989.

289 **The hypothalamic nuclei are connected to the cerebral cortex**
(Kristeva, "Life," 38).

289 **Psycho-phonologists . . . "captives of our vestibules"** (Sollier, *Listening,* 65).

290 **Our "neurobiological networks"** (Kristeva, "Life," 37).

Sonorous Envelopes

Barthes, Roland. "Brecht and Discourse: A Contribution to the Study of Discursivity." In *The Rustle of Language,* translated by Richard Howard, 212–22. New York: Farrar, Straus, and Giroux, 1986.

———. "Deliberation." In *The Rustle of Language,* translated by Richard Howard, 359–73. New York: Farrar, Straus, and Giroux, 1986.

———. "On Leaving the Movie Theater." In *The Rustle of Language*, translated by Richard Howard, 345–49. New York: Farrar, Straus, and Giroux, 1986.

Deleuze, Gilles, and Claire Parnet. "E as in Enfance." In *Gilles Deleuze: From A to Z.* Directed by Pierre-André Boutang. Translated by Charles J. Stivale. Los Angeles: Semiotext(e), 2012. DVD.

Gass, William H. "On Talking to Oneself." In *Habitations of the Word: Essays,* 206–16. New York: Simon and Shuster, 1985.

Klibansky, Raymond, Erwin Panofsky and Fritz Saxl. *Saturn and Melancholy: Studies in the History of Natural Philosophy, Religion and Art.* London: Thomas Nelson and Sons, Ltd., 1964.

LaBelle, Brandon. "Inner Voice, Self Talk." In *Lexicon of the Mouth: Poetics and Politics of Voice and the Oral Imaginary.* New York: Bloomsbury, 2014.

Riley, Denise. "'A Voice without a Mouth': Inner Speech." *Qui Parle* 14 (Spring–Summer 2004): 57–104.

Small, Judy Jo. *Positive as Sound: Emily Dickinson's Rhyme.* Athens: University of Georgia Press, 1990.

301 **"Again, after overcast days"** (Barthes, "Deliberation," 363).

303 **But the impetus for such creations is to be found in the human voice.** From Storr's classic study, *Music and the Mind,* to contemporary

philosopher Peter Sloterdijk's *Bubbles*, we find: "No matter how important lexico-grammatical meaning eventually becomes, the human brain
is first organized or programmed to respond to emotional/intonational
aspects of the human voice" (Storr, quoting Ellen Disanayake in *Music*,
8n15). Sloterdijk describes audible sounds experienced by a fetus in the
womb as "something beyond itself to which it is nevertheless related"
(Sloterdijk, "Siren," 504).

303   **It is different and the same as the voice that I think with.** See Denise
Riley's redressing of the underexplored nature of inner voice, and, via
philosophy of language, its difference from inner thought. Riley asks,
"On what . . . does my conviction of the tonality of my inward voice depend?" (59) and observes: "Even my daily and amiably prosaic mutter-
ings to myself tend to be polyvocal" (71). See also Brandon LaBelle who
asks, "What type of speaking actually occurs within this auditorium of
the 'not quite inner'?" (88), as he aims to reattach the inner voice to the
physical mouth.

305   **"Act in such a way that your own mood could at all times be a reasonable standard"** (Sloterdijk, "Siren," 504).

305   **Édith Lecourt . . . gives us the felicitous image of the baby** (Lecourt,
"Musical," 211).

306   **Édith Lecourt describes a process** (Lecourt, "Musical," 222).

306   **"If the mouth provides the first experience, brief and vivid"** (Anzieu,
Skin, 36).

306   **The poetry of these theories is enticing as when human beings are
understood to contain the qualities of the stars** (Klibansky, Panofsky,
and Saxl, *Saturn*, 264).

310   **"When all escape routes are blocked"** (Kristeva, "Life," 34).

313   **Farinelli . . . Deleuze . . . Dickinson** (Barthes, "On Leaving," 345;
Deleuze and Parnet, *Gilles*; Small, *Positive*, 49). Deleuze refers to his
life-altering meetings by the seaside in Deauville with Pierre Halwachs
who introduced him to the writing of André Gide, Anatole France, and
Charles Baudelaire.

PLAYLIST OF MUSIC
OR SOUND WORKS

(With Links to YouTube Recordings)

"On the Street Where You Live," as sung/performed by:

Vic Damone: https://www.youtube.com/watch?v=dJ4yJM-vquw
Doris Day: https://www.youtube.com/watch?v=ibwzWNbeIB0
Placido Domingo: https://www.youtube.com/watch?v=RBLP24ly9Pk
Eddie Fisher: https://www.youtube.com/watch?v=E94zn6W24Yg
Mario Lanza: https://www.youtube.com/watch?v=8JX8hHcASnw
Peggy Lee: https://www.youtube.com/watch?v=l4JWr1Icffw
Dean Martin: https://www.youtube.com/watch?v=UZz8EzQG5PI
Willie Nelson: https://www.youtube.com/watch?v=RPKHAsDfy7Q

Janet Paige and Fred Astaire, "Stereophonic Sound," *Silk Stockings*: https://www
.youtube.com/watch?v=X0kHKijb8jI
Mitzi Gaynor, "I'm Gonna Wash That Man Right Outta My Hair," *South Pacific*:
https://www.youtube.com/watch?v=Zzu8ZxBHMWk
Darla Hood, "I'm in the Mood for Love," *Our Gang*, 1936: https://www.youtube
.com/watch?v=5QWPi8MtaQk
Judy Garland, "Stormy Weather," *The Judy Garland Show*: https://www.youtube
.com/watch?v=7ioDJZpKCEE
Judy Garland, "A Foggy Day," *The Judy Garland Show*: https://www.youtube.com
/watch?v=Xut7Cpbq4ek
Judy Garland, "Zing Went the Strings of My Heart," in the film, *Listen, Darling*,
1938: https://www.youtube.com/watch?v=uuxyNym5zpk

Sergei Prokofiev, *Romeo and Juliet*: https://www.youtube.com/watch?v=DXyv4SZmKyY

Sergei Rachmaninoff, Piano Concerto No. 2: https://www.youtube.com/watch?v=x8l37utZxMQ

"Both Sides Now" as sung/performed by:

> Joni Mitchell, original studio version, 1969: https://www.youtube.com/watch?v=Pbn6a0AFfnM
>
> Judy Collins, original hit version, 1967: https://www.youtube.com/watch?v=A7Xm30heHms
>
> Bing Crosby: https://www.youtube.com/watch?v=eeKMjg9mcuI
>
> Robert Goulet: https://www.youtube.com/watch?v=0r74ZXTXW8Y
>
> Marie Laforêt, "Je N'ai Rien Appris": https://www.youtube.com/watch?v=94GnxI9NvLw
>
> Leonard Nimoy: https://www.youtube.com/watch?v=_O5nXZ6tmGg
>
> Frank Sinatra: https://www.youtube.com/watch?v=8DrRnI-1Ssg

Tibetan Gongs/Gong Bath Meditation: https://www.youtube.com/watch?v=DHXgiZqSoTg

Sicilian Street Vendors ("A Cry from the Streets"): https://www.youtube.com/watch?v=N0bU-lQvVJo

Sicilian Street Vendors ("Street Vendor in Palermo Selling Fish in a Three-Wheeled Vehicle"): https://www.youtube.com/watch?v=crh6TkMOrik

Jimi Hendrix, "Manic Depression": https://www.youtube.com/watch?v=ppnwGfS7SXg

"Sentimental Journey," the Les Brown Orchestra and Doris Day: https://www.youtube.com/watch?v=PUw125JMVFI

Brahms's Waltzes, Opus 39: https://www.youtube.com/watch?v=MLftlkbqtTY

"White Coral Bells" (in rounds): https://www.youtube.com/watch?v=JE5QrWks1e8

Mark Whittle, "Big Bang Acoustics: Sounds from the Newborn Universe": http://people.virginia.edu/~dmw8f/BBA_web/index_frames.html

Steve Howe, Yes, "Mood for a Day": https://www.youtube.com/watch?v=JE5QrWks1e8

Yes, "Roundabout": https://www.youtube.com/watch?v=-Tdu4uKSZ3M

Leontyne Price singing "Pace, Pace Mio Dio": https://www.youtube.com/watch?v=eGMaoG03fQU

Enrico Caruso and Tita Ruffo singing "Si, Pel Ciel," from Verdi's *Otello*: https://www.youtube.com/watch?v=hMySAq_tiVs&list=RDhMySAq_tiVs#t=0

Enrico Caruso singing "Una Furtiva Lagrima": https://www.youtube.com/watch?v
=Miwejo0mgok&list=RDhMySAq_tiVs&index=4

Stephen Vitiello, "A Bell for Every Minute": https://www.youtube.com/watch?v
=8nymgN5MuDI

Stephen Vitiello, "Intimate Listening": https://www.youtube.com/watch?v
=mTzmJRg8Yyo

PHOTO CREDITS AND
CONTENT DESCRIPTIONS

Plates in "A Mood Room Gallery"

Plate 1. © Rosamond Purcell, bird specimens, yellow, with tags and white envelopes, L. C. Bates Museum, Hinckley, ME.

Plate 2. © Rosamond Purcell, Hemingway blue marlin, L. C. Bates Museum, Hinckley, ME.

Plate 3. © Rosamond Purcell, Mammal Room, L. C. Bates Museum, Hinckley, ME.

Plate 4. © Rosamond Purcell, Hubbard Bird Room with double-wattled cassowary set above cases, L. C. Bates Museum, Hinckley, ME.

Plate 5. © Rosamond Purcell, Victorian bird menagerie display with snowy egret at *center*, L. C. Bates Museum, Hinckley, ME.

Plate 6. © Rosamond Purcell, pileated woodpecker, Charles D. Hubbard diorama, L. C. Bates Museum, Hinckley, ME.

Plate 7. Charles Hubbard, *Owls in Tree*, Charlie Dannie Hubbard, childhood sketchbook, watercolor and pencil, Edith B. Nettleton Historical Room, Guilford Free Library, Guilford, CT.

Plate 8. © Rosamond Purcell, barred owl (*strixvaria*) and snowy owl (*nyctea scandiaca*), Acadia National Park, Charles D. Hubbard diorama, L. C. Bates Museum, Hinckley, ME.

Figures

Figure 1. The author as a child, May Day, circa 1963, photograph, Blessed Virgin Mary School, Darby, PA.

Figure 2. Charles E. Burchfield (1893–1967), *Clearing Sky*, 1920, watercolor on paper, 19¼ × 26½ inches, Burchfield Penney Art Center, Courtesy of the Charles E. Burchfield Foundation ©, Buffalo, NY.

Figure 3. Charles E. Burchfield (1893–1967), *Red Birds and Beech Trees*, 1924, wallpaper from M. H. Birge and Sons Company Pattern 2922, 25¼ × 19 inches, Burchfield Penney Art Center, Courtesy of the Charles E. Burchfield Foundation ©, Buffalo, NY.

Figure 4. Charles E. Burchfield (1893–1967), *Early Spring*, 1966–67, watercolor and charcoal on paper, 37⅛ × 42¼ inches, Burchfield Penney Art Center, Gift of Charles Rand Penney, 1994. Courtesy of the Charles E. Burchfield Foundation ©, Buffalo, NY.

Figure 5. Charles E. Burchfield (1893–1967), *Afterglow*, 1916, watercolor with graphite on paper,19⅜ × 14 inches, Burchfield Penney Art Center, Gift of Tony Sisti, 1979. Courtesy of the Charles E. Burchfield Foundation ©, Buffalo, NY.

Figure 6. Kolya Markov-Riss, circa 2007, photograph, screaming in the garden, Providence, RI.

Figure 7. © Karen Carr, *Girl Playing Mandolin*, circa 1970, figurine, Darby, PA.

Figure 8. Russ and Wallace Berrie Company, "Good-bye Cruel World," 1970, figurine.

Figure 9. Caeli Carr-Potter, childhood rendering of the adults in her midst, circa 2007.

Figure 10. Florence Thomas at work on View-Master set, *Sword in the Stone*, Courtesy Wolfgang Sell.

Figure 11. Shakespeare with moosehead, plaster and taxidermy mount, L. C. Bates Museum, Hinckley, ME.

Figure 12. Barney the bison, taxidermy mount, L. C. Bates Museum, Hinckley, ME.

Figure 13. River otter, *Lutra canadensis* (the otter shares this space with a marten and a mink), Charles D. Hubbard diorama, L. C. Bates Museum, Hinckley, ME.

Figure 14. Red squirrel, detail from the eastern gray squirrel, long-tailed weasel, ermine or short-tailed weasel display, Charles D. Hubbard diorama, L. C. Bates Museum, Hinckley, ME.

Figure 15. Hairy woodpecker (*Picoides villosus*) and yellow-bellied sapsucker (*Sphyrapicus varius*), Charles D. Hubbard diorama, L. C. Bates Museum, Hinckley, ME.

Figure 16. Charles D. Hubbard, the Good Will roundel, 2nd edition, 1917, L. C. Bates Museum, Hinckley, ME.

Figure 17. One of four smaller "museums" at Good Will Hinckley, Granite House a.k.a. Rock House designed by Charles D. Hubbard as it appeared in the 1930s. The millstones were transported from Guilford, Connecticut. Four large and eight smaller murals once hung in the house. A rock garden, also designed by Hubbard, graced the adjacent landscape. Courtesy of L. C. Bates Museum, Hinckley, ME.

Figure 18. Charles D. Hubbard, sample hand-lettering page from Hubbard's *Camp-*

ing in the New England Mountains (Manchester, ME: Falmouth Publishing House, 1952).

Figure 19. American bald eagle exhibit, Leonard C. Sanford Hall of North American Birds, Image #1117, American Museum of Natural History Library, New York, NY.

Figure 20. American bald eagle (*Heliaectue leucocephalus*), Acadia National Park, Charles D. Hubbard diorama, L. C. Bates Museum, Hinckley, ME.

Figure 21. Waterfowl exhibit, Leonard C. Sanford Hall of North American Birds, Image #330321, American Museum of Natural History Library, New York, NY.

Figure 22. Waterfowl: herring gull (*Larus argentatus*), laughing gull (*Larus atticilla*), common tern (*Sterna Hirundo*), Acadia National Park, Charles D. Hubbard diorama, L. C. Bates Museum, Hinckley, ME.

Figure 23. Northern goshawk (*Accipiter gentilis*), Charles D. Hubbard diorama, L. C. Bates Museum, Hinckley, ME.

Figure 24. Charles D. Hubbard, *Ben Mason*, oil on canvas, 17¼ × 13¼ inches, L. C. Bates Museum, Hinckley, ME.

Figure 25. Charles D. Hubbard, Good Will's totem, the beaver roundel, 1924, pen and ink, L. C. Bates Museum, Hinckley, ME.

Figure 26. Walter Reeves appears *second from left*, arm akimbo, and propped on chair. The photo pictures the ten boys who inhabited Good Will Cottage during its first winter, February 1890, alongside Hinckley's first manager, Ernest Ham. Courtesy of L. C. Bates Museum in Hinckley, ME.

Figure 27. Charles D. Hubbard, *Imagination*, 1929–30, 62¾ × 31 inches, East Haven Public Library, East Haven, CT.

Figure 28. Charles D. Hubbard, *Research*, 1929–30, 62¾ x 31 inches, East Haven Public Library, East Haven, CT.

Figure 29. Florence Thomas, *Alice in Wonderland*, plaster bas-relief, 52 × 38 × 2 inches, Beverly Cleary Children's Room, Multnomah County Library, Portland, OR.

Figure 30. Florence Thomas, flower face detail from *Alice in Wonderland*, plaster bas-relief, 52 × 38 × 2 inches, Beverly Cleary Children's Room, Multnomah County Library, Portland, OR.

Figure 31. Leonard Weisgard, illustrated page, in Margaret Wise Brown, *The Noisy Book* (New York: Harper and Row, 1939). Facing page reads: "It was a BABY DOLL. And they gave the baby doll to Muffin for his very own."

Figure 32. Charles B. Shaw, "The fathers were coming home," illustrated page, in Margaret Wise Brown's *The Winter Noisy Book* (New York: Harper and Row, 1947).

Figure 33. Charles B. Shaw, illustrated page, in Margaret Wise Brown's *The Winter*

Noisy Book (New York: Harper and Row, 1947). Facing page reads: "It was a great fire roaring in the fireplace. A big, warm, hot, yellow, roaring fire. And everybody said, 'here is the little dog Muffin. Let him come and warm his nose at the fire.' And Muffin lay down by the fire and listened to the people crack nuts and eat apples and pop popcorn. Someone was eating celery."

Figure 34. Charles D. Hubbard, circa 1883, childhood sketchbook, watercolor and pencil, Edith B. Nettleton Historical Room, Guilford Free Library, Guilford, CT.

Figure 35. Charles D. Hubbard, "Story of the Blind Man," first page, circa 1883, childhood sketchbook, watercolor and pencil, Edith B. Nettleton Historical Room, Guilford Free Library, Guilford, CT.

Figure 36. Charles D. Hubbard, "Story of the Blind Man," second page, circa 1883, childhood sketchbook, watercolor and pencil, Edith B. Nettleton Historical Room, Guilford Free Library, Guilford, CT.

Figure 37. Charles D. Hubbard, *Figures in Orange*, circa 1883, childhood sketchbook, watercolor and pencil, Edith B. Nettleton Historical Room, Guilford Free Library, Guilford, CT.

Figure 38. Charles D. Hubbard, *Bird Landing on Museum*, circa 1883, childhood sketchbook, watercolor and pencil, Edith B. Nettleton Historical Room, Guilford Free Library, Guilford, CT.

Figure 39. Charles D. Hubbard, random illustrations, circa 1883, childhood sketchbook, frontispiece, Edith B. Nettleton Historical Room, Guilford Free Library, Guilford, CT.

Figure 40. Charles D. Hubbard, commercial illustrations (including depictions of a snowshoe, letter *T* with owl, letter *M* with rabbit, Indian feathers on peace pipe, etc.), pen and ink, Guilford Keeping Society Library Collection, Guilford, CT.

Figure 41. Charles D. Hubbard, commonplace book, single page with entries from Friday, July 27, 1928, through Saturday, August 11, 1928, Guilford Keeping Society Library Collection, Guilford, CT.

Figure 42. Charles D. Hubbard, "The people the ways the atmosphere of the olden days," calligraphic emblem as frontispiece for Hubbard's *An Old New England Village* (Manchester, ME: Falmouth Publishing House, 1947).

Part-Opening Illustrations

"Elements": "One of Diamonds." Playing Cards: Engineering. n.d. Engraving. Science, Industry, & Business Library, The New York Public Library, Astor, Lenox, and Tilden Foundations.

"Charts": "Knave of Spades." Playing Cards: Engineering. n.d. Engraving. Science, Industry, & Business Library, The New York Public Library, Astor, Lenox, and Tilden Foundations.

"Rooms": "The Eight of Diamonds." Playing Cards: Engineering. n.d. Engraving. Science, Industry, & Business Library, The New York Public Library, Astor, Lenox, and Tilden Foundations.

"Vibes": "The Five of Hearts." Playing Cards: Engineering. n.d. Engraving. Science, Industry, & Business Library, The New York Public Library, Astor, Lenox, and Tilden Foundations.

INDEX